Space-Time Physics: The Revolution in Physics For the New Millennium

By

Jesús Parrilla-Calderón

This book is a work of non-fiction. Names and places have been changed to protect the privacy of all individuals. The events and situations are true.

ISBN: 1-4107-9631-0 (e-book)
ISBN: 1-4107-9632-9 (Paperback)
ISBN: 1-4107-9633-7 (Dust Jacket)

Library of Congress Control Number: 2003096864

This book is printed on acid free paper.

Printed in the United States of America
Bloomington, IN

1stBooks – rev. 01/07/04

PREFACE

The ideas presented in this book are the result of about 25 years of meditations and efforts to develop a consistent theoretical framework from which all physical manifestations of the universe can be developed or at least inferred. I have made several attempts to publish articles about my ideas about physics and all have resulted "speculative, absurd, uncomprehensible, preposterous" or even "senseless" to referees. Therefore, I am writing for educated lay persons and physicists who dare pay attention to such speculative, absurd, uncomprehensible, preposterous or senseless ideas about physics.

I do not pretend that this book will begin a new revolution in physics, but I write it with the objective that the ideas presented in it will find an open and illuminating mind with the capacity to subject those basic ideas to a rigorous mathematical analysis, most probably twistors mathematics, bringing up a new breakthrough in physics. Albert Einstein has been quoted as saying

There is no logical path to the laws of the cosmos. Only intuition based on a sympathetic understanding of experience can reach them.

I claim that I have had the intuition and the understanding of experience, but I do not manage twistors mathematics that I know that is needed. Therefore, this book pretends to let the ideas be known, so that those who read them will have the basic ideas upon which a future breakthrough in physics may ensue; and some with the physics and mathematical tools and insight, and the disposition to spend time and effort on speculative ideas, may produce that breakthrough.

All basic physics ideas are simple and amenable to the educated lay person, and to all physicists. Those basic physics ideas presented here are not an exception. Thus, this book may be very interesting to all educated persons, but very particularly, to all physicist who care about daring and potentially revolutionary, speculative ideas in physics. So, if

you are one of those daring and/or curious ones, welcome.
Go on reading.

INTRODUCTION

POSTULATES FOR A NEW SPACE-TIME

Physicists deal with four fundamental observables of nature; mass, space, time and electric charge. The last one, charge, is actually an interaction of material bodies. As a consequence, one may reduce the fundamental observables to three; mass, space and time.

Mass, space and time are unseparable quantities: For man to acertaine the existance of space he must reallice the posibility of displacement; and displacement means a material body to be at a given place a given time, and at another place later; thus, within the idea of space, mass and time are implicit: And to acertaine the pasage of time a displacement of material bodies is needed. -Be it motion of the Sun, stars, planets, moon, sand, a pendulum, a spring, atoms in a molecule or photons-motion of material boddies are needed to determine time. (I challenge the reader to think of a way in which time can be determined without matter displacement): Matter (mass) can exist only in space and time.

If mass, space and time are unseparable, is'nt it logical to think that they are manifestations of one fundamental thing. My answer is yes, and I call that thing **SPACE-TIME.**

The theory presented here is based on a general conclusion that **motion, particularly rotation, is the most fundamental and universal physical phenomenon in nature**: and that there exist a universal "rotation", **spin,** which for hevenly bodies is a normal rotation in the normal four dimensional physics space, one dimension of time and three of space, known as **four-space;** but for subatomic particles other non convensional spaces must be used in order to explain the same phenomenon. The space used is **a function space,** that is, a space where every dimension is on itself a function of the physical **four-space.** A logical alternative is to explore a more general physical **space-time** (We shall call it **SPACE-TIME**) where non conventional rotations can occur. After long and strenuous analysis of the situation the author has come to the conclusion that giving space and time equal standing seems to be the solution.

In what fallows I present ten postulates which are the basis for further development of the theory. As most theoretical physics postulates, at first glance, they may seem rare or even absurd. But through the different chapters of the book the reader will find that they are based upon well known physics.

POSTULATES

1- The only physically, fundamental thing in nature is **SPACE-TIME.**

2- Space and time are **equivalent** manifestations of **SPACE-TIME.**

3- If there is (or is not) a possibility of a **return** in space, then there is (or is not) that possibility in time.

4- In our material world space has three **clear and evident** dimensions, therefore, fallowing postulate #2, time has three dimensions.

5- In our material world there seems to exist only one dimension of time. We call this a **preferred direction (or dimension) of time**. Then, by symmetry, there should be a **preferred direction of space.**

6- Space rotations exist in all manifestations of the material world, then time rotations should also exist.

7- In our material world **space-time** seems to be three dimensional in space and one preferred dimension in time. Then, by symmetry, there should exist another manifestation of the universe in which **space-time** must seem to be three dimensional in time and one preferred dimension in space.

8- By postulate # 7, physical **SPACE-TIME** consists of our material **space-time** that has three clear and evident dimensions of space and three of time, one of which is preferred; and a **dual space-time** where time behaves like space, and space like time, that has three clear and evident dimensions of time and three of space, one of which is preferred. Therefore, the physical **SPACE-TIME** has twelve dimensions.

9- In our material world, motion is defined as the variation of a space dimension on a time dimension, then a **generalized motion** is a variation of any one dimension on another.

10-All manifestations of the universe can be represented as generalized motion on the twelve dimensional **SPACE-TIME; space-time and its dual space-time.**

Of the above ten postulates the first and the last are the strongest and basic ones:

1- The only physically, fundamental thing in nature is SPACE-TIME.

10- Al manifestations of the universe can be represented as generalized motion on a twelve dimensional SPACE-TIME.

Postulate #2 puts <u>time</u> on equal footing as <u>space</u>: Postulates #3 through #8 may be said to be consequences of #2, but they include additional experiences or viewpoints that give them standing as postulates: Postulate #9 is actually a generalized definition of motion, which is a necessity once we give time and space equal standing. Thus, postulates #2 through #9 are like links of a chain that connect #1 to #10.

It has been said that time and space are on equal footing since the introduction by Minkowski of the four dimensional physical SPACE. But, since in that SPACE time was kept at only one dimension, then it is not equivalent to space. To place <u>time</u> and <u>space</u> on equal footing the twelve dimensional **SPACE-TIME** we have postulated seems to be a necessity.

Postulate #5 puts the law of conservation of angular momentum at stake. But it should come as no surprise, for the direction of a gravitational field at any locality is a **locally preferred direction:** For, if a spinning body is placed on the field, held through a fixed point out of its center of mass and free to rotate about that point, it will either spin alined with the field or precess about the field direction. (At any given point in space there may be different preferred space direction for corresponding different fields.) Analyzing the spin and precession of rotating bodies, in 1974, I inferred that there should exist a cosmologically preferred direction of space. And in 1997 Nodland and Raldstone[1] claimed to have found an anisotropy in electromagnetic wave propagation over cosmological distances which suggest the existence of a

---

[1] Borge Nodland and John P. Ralston, Physical Review Letters, Vol. 78, Num. 16, April 1997.

preferred direction of space. Thus, it may be necessary to broaden the scope of the law of conservation of angular momentum to include the existence of a cosmologically preferred space direction.

What we pretend with postulate #10 is that all physics is actually **kinematics.** That is, that mass, electric charge, and all fundamental forces are manifestations of motion on a generalized space. Therefore, in the following chapter we shall present cases of well established physics in which space or motion play an active roll.

CHAPTER I

SPACE AND TIME AS THE FUNDAMENTAL THING

In this chapter we present evidence that many of the fundamental physical phenomena with which physicists normally deal can be represented by a particular property of space or by some kind of motion. This will serve as basis for a generalization into postulates #1 and #10.

1.1   The contact force

We begin this chapter with a trivial case in which motion produces a force, namely the contact force. This force is generated when one body tries to occupy the space that another body is occupying. In a collision, for instance, two bodies are changing position in space and at a given instant they try to occupy the same position and a force generates. In a sense, the force is generated by their motion.

I.2   The concept of a force field

A force field-gravitational, electrical or magnetic-has been, for long, used to represent the effect of material bodies (or electrical charges) upon other bodies (or charges). The gravitational effect of a mass, m, upon another mass, M, is substituted by a property that m produces through all space, called a **gravitational field**. Knowing the field it produces we can ignore m completely. That is, the effect of m upon any other mass of the universe is represented by a property that it gives to space, the gravitational field. Identical situation exists with a static electrical charge. Its effects can be represented by the electrostatic field it produces.

The reality of a field being a property of the space by itself can be realized when we consider a collision of two particles: Let's consider two positively charged particles colliding. For example, a proton colliding with an alpha particle. The interaction is between the spaces around the particles. The space around the proton collides with the space around the alpha particle and in the collision the material particles need never come in contact. Even in a head on collision, in which both particles reverse their original direction of motion, material contact need never occur. (A visual model for this collision is that the space between the two particles acts like a spring being compressed by the two particles moving toward each other) In a case of an electron colliding with a proton the situation is more dramatic: The space around the electron collides with the space around the proton and both particles may deviate from their original path in space; or in the collision the two spaces may alter completely and become a totally different space configuration known as a neutron and neutrino, which is not the sum of an electron and a proton.

The case of a moving electric charge (electric current) is a well known example of an effect produced by a motion. If a magnetic dipole (a magnet) is in the vicinity of a moving electric charge, the dipole experiences a torque (Except if the motion of the charge is along the line defined by the axis of the dipole). If the charge doesn't move, there is no torque. Conversely, if a magnetic dipole rotates or moves in any way, any electric charge in its vicinity, out of the line defined by the axis of the dipole, will experience a force. In both cases motion originates a force. These are cases known for more than a century in which forces are generated by motion. They are the basis for the electric motor and the electric generator. **Motion of a charge generates a field (and a force or torque), and motion of charge through a field generates a force on the charge.** In these cases, clearly, motion generates forces.

Now, consider two identical, coaxial, circular wires on planes separated by a small distance. If charges are forced to circulate in both circles, in the same sense, the circles will attract each other: if forced to circulate in opposite senses they will repeal. This is a consequence of the well known fact that parallel electrical currents attract and antiparallel repeal. Or, equivalently, opposite magnetic poles attract and equal magnetic poles repeal. In these cases rotation of electrical charges generates a force.

The change in motion of an electric charge, an acceleration, produces the interesting effect of energy being sent out, irradiated, throughout space. This is the well known mechanism for producing electromagnetic waves, such as radio, radar and microwaves, as well as X rays. Thus, here we have a change in motion producing electromagnetic radiation. One may reasonably claim that in this case the ultimate cause of the radiation is the force that causes the change in motion on the electric charge. This brings us back to the classical physics problem of causality in the equations of dynamics. That is, are forces the causes of accelerations as was classically given for granted? Or, contrary to this, can one say that accelerations are the causes of forces?

To analyze this classical causality situation let us consider the gravitational interaction of two isolated bodies in the universe. The physical phenomenon that an observer on one of the bodies (or an outside observer) can "see", and measure, is the change in distance on different intervals of time; and from the data he can calculate the relative acceleration of both bodies; and from the acceleration he can calculate the force only if he knows the mass of one of the bodies. The only way to determine the force is by measuring positions, from which acceleration is calculated and from the acceleration the force is calculated. One may say that in this case the acceleration is the observable phenomenon. The force is a "man's mathematical construct" to explain and quantify the observed phenomenon.

Contact forces also are observed or measured only by means of a change in position and time. For example, a spring that **was** (time implied) of a given length is longer now under the action of some external elements on its extremes. The observable phenomenon is that the spring has changed its length; we then infer that a force has acted on its extremes.

The third law of motion (Newton's third law) asserts that *if **one** body exerts a force on a **second** body, then the **second** exerts an equal and opposite force on the **one**.* This is usually called the law of action and reaction. But it is not possible to identify clearly which is the action and which the reaction. What the law means, without ambiguity, is that **forces exist always in couples:** If body **A** pushes or pulls upon body **B**, then **B** pushes or pulls upon **A** with equal force in the opposite direction. For analysis one can select the action force at will; the other will be the reaction. For instance, let's say that the force of **A** upon **B** is the action. Say also that it is the cause of an acceleration on **B.** Then, one may say that the acceleration on **B** causes the reaction force upon **A.** Therefore, the classical cause and effect relationship between force and acceleration is not in solid grounds. It can be reversed without any effect on the theory. The case of a body moving on a circle, tied to a string, is a clear and simple example: The force of the string upon the body causes an acceleration of the body; the acceleration of the body causes a force on the string.

Consider a hypothetical case of a community that was born and developed inside a very large cylinder of radius R, in outer space, rotating with angular velocity w, relative to the rest of the universe. They do not know that they are inside a rotating cylinder. They fill that their bodies and all other objects are pulled "down". (Toward the wall of the cylinder due to the well known centrifugal force). If they develop physics as was developed here on earth, working on distances much smaller than R, the radius of the cylinder, they will conclude that in their world all bodies "fall" with a constant acceleration. They may call it g, a constant (We know that the acceleration is w^2R, a constant). Since the distances they work with are much smaller than R, they will conclude that their world is flat; they may develop projectile motion exactly as we have developed it in our "flat" surface. They may discover that at high altitudes g varies a little, (as here on earth) and may even discover the law by which it varies. Experimentally, they may discover directional anomalies in their projectile motion (Coriolis effect). And they may develop directional corrections for their long range projectiles.

The reader may have noticed that the above paragraph shows that the centrifugal acceleration is nearly identical to a gravitational acceleration. In fact, if the environment within the hypothetical cylinder is identical to one on the earth surface, and a scientist in an unconscious state were transferred from the earth to the cylinder, awakening, he will not notice any difference and will conclude that he is still within the earth gravitational field. (From general relativity we learn that a gravitational field is equivalent to an acceleration)

For that community living all their lives within the rotating cylinder in outer space, there exits a force field in their space just like there is one in our earth surface. For them, the centrifugal force is as real as the gravitational force is for us. From the point of reference of outside observers who know that the cylinder is rotating relative to the rest of the universe an acceleration is causing that force.

Generalizing: **Accelerations generate forces**. For particular cases of accelerated bodies, the classical third law

of motion can be stated as fallows: *An acceleration,* ***a***
(vector) of any body of mass M, causes on a second body, a
*force **F**, opposite in direction to **a** and of magnitude F= Ma.*
(Vectorially, the equation would be ***F= -Ma***). Notice that,
by the principle of superposition of forces, the body upon
which ***F*** acts may be composed of many bodies or particles.
For example, the Earth is accelerated relative to the center
of mass of the rest of the solar system. That acceleration can
be said to cause a force upon all the rest of the members of
the solar system as a body. But that force is composed of
the sum of individual forces upon the Sun, the Moon, and
all other members of the system.

1.3   Force generated by two perpendicular rotations

In the hypothetical cylinder example discussed above
we have presented a force generated by a rotation, a
centrifugal force, known since the times of Newton. In this
section we present a new force generated by two
perpendicular rotations. That is, that two perpendicular
rotations simultaneously on a rigid body generate a force.

This will be one step further in the development of our theory that forces are generated by motion.

1.3.1 Some preliminary facts

Figure 1 Represents a wheel rotating about its principal axis with angular velocity w_1. An extreme of a rigid axis is connected to a pivot point at the origin of coordinates, free to rotate about any of the coordinate axes. At the beginning the axis is held horizontally while the wheel rotates about its axis with angular velocity w_1 and angular momentum L_1. If the system is then forced to rotate about the Z axis with angular velocity w_2 and angular momentum L_2, its center of mass may rise to a point that will depend on the magnitude of the angular velocities; Or a force acting downward on the axis, of magnitude $\mathbf{F_{cc}}$ (this will be developed below) applied at the radius of gyration will be necessary to prevent it from rising.

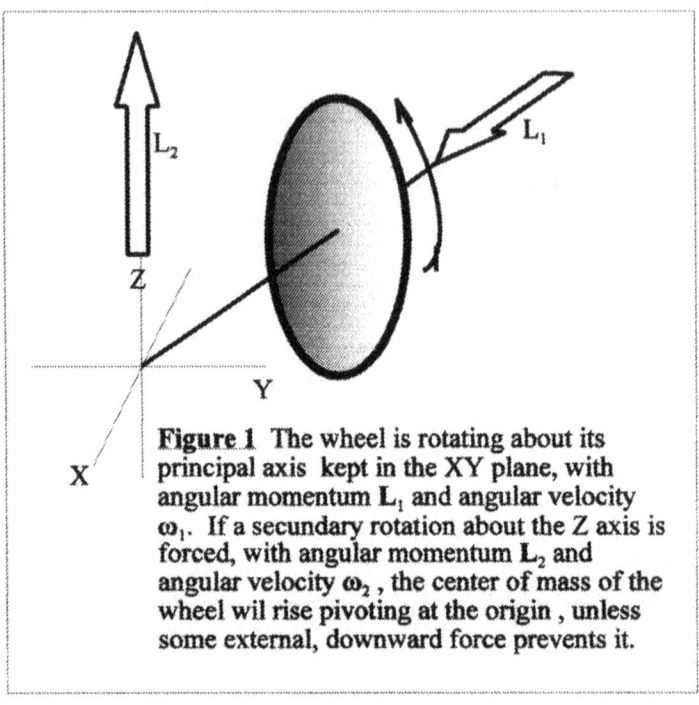

Figure 1 The wheel is rotating about its principal axis kept in the XY plane, with angular momentum L_1 and angular velocity ω_1. If a secundary rotation about the Z axis is forced, with angular momentum L_2 and angular velocity ω_2 , the center of mass of the wheel wil rise pivoting at the origin , unless some external, downward force prevents it.

On June 10, 1980 I submitted a work with the title **Rotations as Generators of Forces** to the American Journal of Physics for publication. (Copy of this article is included as APPENDIX I) I included the force F_{cc} that I am presenting here with an equivalent figure. They rejected the paper with the fallowing commentary of a referee (among others) **"On page 10 an apparatus is briefly discussed in which F_{cc} can be measured. If the author has (as he claims) detected F_{cc} to be non-zero, it would be a great**

boon to mankind. This measurement of levitation, or its expectation, demonstrates the incorrectness of his arguments." This commentary reflected a misunderstanding of the work, for it does not imply levitation. The rise occurs because the system pivots on the fixed pivot point. It is analogous to the case of an athlete who jumps using his motion and a rod to pivot on a fixed point on the earth. However, recognizing the existence of this force may lead to interesting applications. Very large mechanical advantages may be obtained using two perpendicular rotations on a rigid body.

Reacting to the referee's commentary cited above I built a system where the effect of the force was pictured. I prepared a paper for publication. Stimulated by the referee's commentary I sent it to Nature, in London, on October 23, 1980. (Copy of the paper and the reply from Nature is included as APPENDIX II) Nature rejected it claiming that it was not of sufficient impact to merit publishing in Nature and recommended that I send it to the American Journal of Physics. I did, and it was rejected with the commentary that there was nothing new in the experiment, that the same had been done at Berckly some years earlier. Strangely, the journal that rejected the first paper with the statement that **"If the author has (as he claims) detected F_{cc} to be non-zero, it would be a great boon to mankind."** rejected a paper where the presence of the force was demonstrated, beyond doubt, experimentally, stating that it was nothing new.

In 1989 Dr. E. H. Puthoff, of the Institute for Advanced Studies at Austin, Texas, visited us at Cayey University College and we discussed these matters. He red my paper about the devise that I had designed, pictures of which I had sent to Nature in London on October 23 1980. He said that Dr. Eric Laithwaite, of London University Imperial College, was doing similar wok and had applied for patent for devises generating forces with this gyroscopic motion. I have a copy of Dr. Laithwaite patent claim and, in fact, in principle, it is the same devise that I had designed a decade before, sent it to Nature for publication and was rejected.

1.3.2 Mathematical derivation of the force generated by two perpendicular rotations.

For the non scientist or mathematician, this sub-section may be mathematically out of reach. However, for the purpose of the lay reader the conclusion of this section is what matters. That is, **two perpendicular rotations generate a force.** Therefore, if you have difficulty with the mathematics, please, skip this sub-section.

Consider the special case in which the wheel of <u>figure 1</u> is rotating with its axis horizontally, with velocity w_1 and simultaneously it is rotating about the Z axis with angular velocity w_2. Referred to the stationary system it has angular momentum (In this section **bold letter** implies vector quantity)

$$\mathbf{L_s} = \mathbf{L_1} + \mathbf{L_2} = I_1\, \mathbf{w_1} + I_2\, \mathbf{w_2} \text{----------------------}(1.1)$$

where the I's are the corresponding moments of inertia. The products of inertia are zero due to the symmetry about the main axis and about any plane passing through the axis.

Since w_1 is changing direction L_1 is changing with time, and L_s also changes. To evaluate the rate of change of vectors referred to different frames we use the equation

$$(dA\,/dt)_s = (dA/dt)_r + w \times A \text{ ----------------------- } (1.2)$$

where $(dA\,/dt)_s$ and $(dA\,/dt)_r$ stand for time-rate of change of vector A relative to the stationary and rotating system respectively. (See any intermediate or advanced text in mechanics. For example, Arthur and Fenster, Mechanics, Rinehart & Winston, 1969, Eq.12.49: Or E. Neal Moore, Theoretical Mechanics, John Wiley & Sons, 1983, Eq. 6.21). w is the angular velocity of the rotating frame of reference, which in this case is w_2 and is a constant. Applying this to Eq. (1.1) we obtain

$$(dL_s\,/dt)_s = (dL_1\,/dt)_r + (dL_2\,/dt)_r + w_2 \times L_1 \text{ ------ } (1.3)$$

because $w_2 \times L_2 = 0$, since w_2 and L_2 are parallel. From the rotating frame of reference L_2 and L_1 are constants. Therefore $(dL_2\,/dt)_r = (dL_1\,/dt)_r = 0$. Thus,

$$(d\mathbf{L}_s /dt)_s = \mathbf{w}_2 \times \mathbf{L}_1 = \mathbf{w}_2 \times I_1 \, \mathbf{w}_1 \text{ ------------------ } (1.4)$$

Newton's second law applied to rotations, $d\mathbf{L}/dt = \text{torque} =$ **N**, in this case leads to

$$\mathbf{N} = \mathbf{w}_2 \times I_1 \, \mathbf{w}_1 \text{ --------------------------------------- } (1.5)$$

Eq. (1.5) means that for the system to have the two rotations $\mathbf{w}_1$ and $\mathbf{w}_2$ a torque **N** must be acting upon it. This equation may be found in many dynamics books. The interpretation that follows is my contribution.

Eq. (1.5) is analogous to the case of a particle tied to a string, moving on a circle. In that case a centripetal force $\mathbf{F}_c$ must act on the particle. And, by the third law of motion, a centrifugal force $-\mathbf{F}_c$ is exerted by the particle on the string. By analogy, then, Eq. (1.5) means that the system exerts a torque $\mathbf{N}_{cc} = -\mathbf{N}$ (I use index cc to associate with double rotations) on whatever is exerting the torque **N** on it. $\mathbf{N}_{cc} = - \mathbf{w}_2 \times I_1 \, \mathbf{w}_1 = I_1 \, \mathbf{w}_1 \times \mathbf{w}_2$

$$N_{cc} = I_1 \, w_1 \times w_2 \text{ --------------------------------- } (1.6)$$

Any torque is said to be produced by a couple. That is, by two forces acting on opposite directions. The system we are analyzing has one fixed point, the pivot point. The system pushes down at that point. Since the torque is produced by the two rotations it is logical to assume that the other force of the couple is acting at a point associated with the rotation. Furthermore, the torque is about the secondary axis (Z axis). Therefore, *the second force of the couple is on the principal axis at a distance associated with the rotation about the secondary axis.* Then, we assume that it acts at the radius of gyration, k, perpendicular to the principal axis. The torque is then, $N_{cc} = \mathbf{k} \times \mathbf{F_{cc}}$ Since $\mathbf{k}$ and $\mathbf{F_{cc,}}$ as well as w_1 and w_2 are perpendicular $N_{cc} = k \, F_{cc}$; $|w_1 \times w_2| = w_1 \, w_2$. Therefore, $k \, F_{cc} = I_1 \, w_1 \, w_2$, or

$$F_{cc} = (I_1/k)w_1 \, w_2 \text{ --------------------------------- } (1.7)$$

By definition of the radius of gyration, $I_2 = mk^2$. Substituting k on Eq. (1.7) we obtain

$$\mathbf{F}_{cc} = (I_1{}^2 m/I_2)^{\frac{1}{2}} w_1\ \mathbf{w}_2 \text{----------------------------} (1.8)$$

We have used vector notation in Eq. (1.8) since from the geometry it is easy to show that $\mathbf{F}_{cc}$ is in the direction of $\mathbf{w}_2$. In APPENDIX I the vector equation is derived directly and in a more general way.

We realize that the force that the system exerts can be assumed to act at any other point on the axis, out of the origin. For, after all, what is unquestionable is the existence of the torque given by equation (1.6). But equation (1.8) is the only one yielding a result that include parameters that strictly relate to the rotations of the system. For instance, we may calculate the force that the system would exert at any distance R from the z axis. This is obtained from the equation $\mathbf{N} = I_1\ \mathbf{w}_1 \times \mathbf{w}_2 = \mathbf{R} \times \mathbf{F}$. For this particular case in which $\mathbf{w}_1$ is perpendicular to $\mathbf{w}_2$, this vector equation leads to the scalar equation $I_1\ w_1 \times w_2 = RF$, from which

$$\mathbf{F} = (I_1/R) \, w_1 \, \mathbf{w}_2 \text{ ------------------------------------ } (1.9)$$

Equation (1.9) shows that the force varies inversely with R. Extremely large forces would be exerted at small distances from the z axis. But only F_{cc} depends exclusively on rotations parameters.

CHAPTER II

MATHEMATICAL EQUIVALENCE OF TIME AND SPACE

In this chapter we establish the mathematical rudiments upon which the theory rests. For the lay readers this may need review of mathematics. But you may pass through this chapter, get the general feeling of it and go on to the next without loss of the line of reasoning of the book.

In Minkowski's space (four-space) time is just the fourth coordinate of the physical space. Coordinates are the time coordinate, $x_4 = ict$, where c is the velocity of light in vacuum and i is the square root of -1; and the regular three space dimensions, say, $x_1, x_2, x_3,$. This conventional notation for the time coordinate allows us to represent the metric of the space, ds^2, in the usual way

$$ds^2 = (dx_1)^2 + (dx_2)^2 + (dx_3)^2 + (dx_4)^2 \text{------------ (2.1)}.$$

Or, considering that $x_4 = ict$, $dx_4 = icdt$ and $(dx_4)^2 = -1c^2dt^2$, Eq. (2.1) becomes

$$ds^2 = (dx_1)^2 + (dx_2)^2 + (dx_3)^2 - c^2dt^2 \text{------------- (2.2)}.$$

By taking the unit of time to be $1/c$ seconds, (This unit of time may be called the light-meter, that is, the time it takes light to travel one meter in vacuum) then the speed of light in vacuum becomes 1, and equation (2.2) can be written as

$$ds^2 = (dx_1)^2 + (dx_2)^2 + (dx_3)^2 - (dt)^2 \text{------------ (2.2-a)}$$

In equation (2.1) the metric, ds^2, has all components positive and is said to have a *signature* **positive definite**: In equation (2.2-a) the *signature* is +,+,+,-. This is the signature of Minkowski space. But nothing precludes us from using a positive definite signature until the actual signature is determined.

In the SIX-SPACE, **X,** that we have postulated,

$$\mathbf{X:} \; x_1, x_2, x_3, x_4, x_5, x_6 \text{ ----------------------------- (2.3)}$$

where the xs are our material world six coordinates. We single out the time coordinate $x_4 = ict$ as the preferred time direction of our material world; and x_5, x_6 are the corresponding hidden time dimensions. The dual SIX-SPACE, **Y,** is represented by the six coordinates

$$\mathbf{Y:} \; y_1, y_2, y_3, y_4, y_5, y_6 \text{ --------------------------- (2.4).}$$

The preferred space coordinate is y_4; and the hidden space coordinates are y_5, y_6.

By postulate #5, **the preferred dimension of time, $x_4 = ict$, becomes the preferred dimension of space, y_4 on the dual space.** This means that the dual space, **Y,** is the **mirror image** of **X.** What is space in **X** reflects as time in **Y:** what is time reflects as space.

$$y_4 = k x_4 \text{---} (2.5)$$

where **k** is a time to space conversion operator. The inverse operator should also exist.

$$C = k^{-1} \text{---} (2.6)$$

so that $C y_4 = k^{-1} k x_4 = x_4$.

Analogously, there should exist corresponding operators for the rest of the six coordinates. The dual space **Y** should be obtained through a lineal transformation from **X.** The transforming matrix is, then, a vector operator, **K,** that transforms space into time and time into space. We call it the **Space-time operator.** The *k* operator of equation (2.5) is an element of *K* corresponding to x_4 and y_4. So, we shall write it as k_4. The vector operator *K* is a diagonal matrix of elements k_1., k_2., k_3., k_4., k_5., k_6. In general, k_i is *the element of K transforming x_i into y_i.*

Equation (2.7) is the matrix representation of the transformation of *X into Y* by the **Space-time operator K.**

$$KX = \begin{pmatrix} K_1 & 0 & 0 & 0 & 0 & 0 \\ 0 & K_2 & 0 & 0 & 0 & 0 \\ 0 & 0 & K_3 & 0 & 0 & 0 \\ 0 & 0 & 0 & K_4 & 0 & 0 \\ 0 & 0 & 0 & 0 & K_5 & 0 \\ 0 & 0 & 0 & 0 & 0 & K_6 \end{pmatrix} \begin{pmatrix} X_1 \\ X_2 \\ X_3 \\ X_4 \\ X_5 \\ X_6 \end{pmatrix} = \begin{pmatrix} K_1 X_1 \\ K_2 X_2 \\ K_3 X_3 \\ K_4 X_4 \\ K_5 X_5 \\ K_6 X_6 \end{pmatrix} = Y ---------(2.7)$$

Mathematically, the operator *K* establishes unequivocal equivalence of space and time. But to establish a physical meaning for a change of space into time, or vice versa, is another story. The explicit physical meaning of the operator *K* must be established. However, to deal with physics of our material world we need only what we have called our material SIX-SPACE **X.** What postulate #10 asserts is that all our material world physics should be represented as generalized motion in **X.** Transformations into **Y** could represent anti-matter, black matter, tachyons as well as other possible phenomena not covered by present-day physics.

CHAPTER III

DIFFERENT MOTIONS ON A SIX-SPACE

3-1   Introduction

We have postulated that physical SPACE-TIME
consists of a six dimensional space and its dual. And that **all
manifestations of the physical world can be represented
as motion on SPACE-TIME.** But motion on a frame of
reference where space and time have equal standing cannot
be only derivatives with respect to time. Any change in a

coordinate, Δx, is motion. In this chapter we analyze the different motions on a six-space.

The space has two sub-spaces, each one of three dimensions, one corresponding to SPACE and another corresponding to TIME. That is,

$$X = S + T \text{---} (3.1)$$

where X is the six dimensional SPACE-TIME; S and T are the three dimensional sub-spaces corresponding to SPACE and TIME, respectively. Also, Minkowski space, M_4, is a subspace of the SIX-SPACE. Therefore we can write $X = M_4 + T_2$, where T_2 represents the two additional dimensions of time that I am proposing.

The first question we must face is, what is the nature of SPACE-TIME? Is it a void? Definitely not! All cosmological theories point to the fact that what we call a vacuum is not a void. In fact, many scientists claim that the zero point energy of quantum mechanics is energy of the vacuum. That is, that vacuum is actually a sea of energy.

Until the beginning of the twentieth century scientists believed that vacuum was actually full of what they called the **ether** in which electromagnetic waves propagate. The fact that the speed of light in vacuum is independent of the relative velocity of source and observer led to abandon the ether theory. This gave the speed of light in vacuum an absolute value. Paradoxically, the theory of relativity starts from the postulate of a non-relative velocity for light and all electromagnetic waves in vacuum. This sets electromagnetic waves aside from all other physical manifestations of the universe. How can this be explained?

3.2   Anti-matter as matter on the dual SPACE-TIME

How can the speed of light in vacuum be non relative? Seeking an answer to this question we may use the metric in vector form, in Minkowski space,

$$\mathbf{ds} = \mathbf{i}\, dx_1 + \mathbf{j}\, dx_2 + \mathbf{k}\, dx_3 + \mathbf{l}\, dx_4 \text{-----------------(3.2)}$$

where (**i, j, k, l**) are orthonormal unit vectors along the corresponding coordinates (x_1, x_2, x_3, x_4). We assume that x_4 is linearly dependent on time. And is the only time dependent variable on the metric. That is, x_4 can be written as $x_4 = a\tau$ where a is a constant and τ is one of the three time dimensions of six space. Consider a light pulse moving through space. If we perform measurements to determine its speed the value obtained is always c, independent of the relative velocity of source and observer. From equation (3.2) we obtain

$$ds/d\tau = \mathbf{l}\, d\,(a\tau)/d\tau = \mathbf{l}\, a \text{ --------------------------}(3.3)$$

We know that the magnitude a = c.; and that the unit vector **l** is rotated 90 degrees from the other SIX-SPACE dimensions. Rotation by 90 degrees is obtained by applying the imaginary number i. That is

$$ds/d\tau = ic \text{ ---------------------------------------} (3.4)$$

This suggests that the speed of light in vacuum, c, is not a change in position with respect to regular (preferred) time, but with respect to a time dimension rotated $90°$ from regular time. This may explain why that speed is nonrelative to us; for we perceive the world from a preferred time viewpoint.

In four-space, M_4, material particles are in motion along the preferred time coordinate as long as they exist, because their time coordinate is changing. This is also true in our six SPACE-TIME, since M_4 is a subspace of X. This implies that there exists an absolute frame of reference for SPACE-TIME. This would agree with the cosmological theory that time and space began with the **big-bang**. We would rather say that SPACE-TIME has an absolute origin at the big-bang.

Relativity theory deals with the perception of the material world from frames of reference in Minkowski space that are in motion relative to the big-bang. It has been established theoretically and experimentally, that in that realm the speed of light in vacuum is a constant, c. But using the result obtained on equation (3.4), namely that the constant c is a velocity with respect to a hidden time dimension, we may set the origin of coordinates so as to make the speed of light in vacuum, that is, the derivative of space coordinates with respect to the preferred time coordinate, equal to zero. Therefore, it is reasonable to assume that in our six SPACE-TIME frame of reference whose origin is at the big-bang, electromagnetic waves, photons, are at rest in M_4, while they may be moving with speed c respect to the other dimensions of time. And every other particle of the universe is moving in M_4 as long as it exists. This suggests that antiparticles exist in the dual SPACE-TIME. If a particle and antiparticle collide they come to rest in M_4, and are converted into photons. They annihilate. Thus, photons belong to both, SPACE-TIME and its dual. This explains why antiphotons do not exist.

Pair production, the process inverse to annihilation, is simply a collision of a particle, moving for ever in SIX-SPACE, with a photon at rest in M_4. This explains why pair production occurs always in the presence of material particles.

The universe is full of photons at rest in M_4. Photon emission is always a collision of ever-moving particles with photons at rest in M_4, producing new photons at rest. Photon absorption is a collision in which a photon at rest in M_4 is absorbed by the particle. The moving particle may be elementary; or composite (atom or molecule). All processes exhibited by matter in space are also exhibited by anti-matter in the dual space.

3.3   Possible motions on SPACE-TIME

In quantum mechanics the state of a particle or system of particles is a vector in a Hilbert Space whose basis functions are functions on M_4. That is, they are functions of x_1, x_2, x_3, x_4, (x, y, z, t). What I am proposing is that those basis functions are functions of six variables, x_1, x_2, x_3, x_4, x_5, x_6, of SIX-SPACE.

Any particle or system of particles in the physical world will be represented by a group of rotations in X. And the state of the system will be a vector in the Hilbert

Space whose substrate is *X*. Motion of the system is any change in its state. In this section we identify the different kinds of motion that are possible in that SPACE-TIME.

3.3.1 Rotations on SIX-SPACE

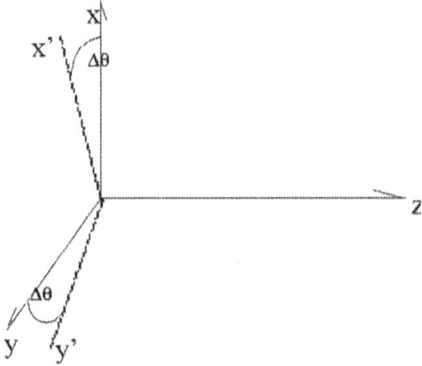

Figure 3.4 A rotation on M_4

The first motion that concerns us is rotation. In M_4, a constant rotation means a two dimensional spacial plane changing with respect to the time dimension about the remaining spacial dimension. This is illustrated in Figure 3.4. The angular velocity is, (Bold letter or arrow over letter means vector)

$$\vec{\omega} = \frac{d\vec{\theta}}{dt} = \vec{k}\frac{d\theta}{dt} \quad -------------- (3.5)$$

where **k** is a unit vector along the z dimension.

In equation (3.5) θ represents the x,y plane, z the axis and t the independent variable. That is, in a four dimensional M_4, x,y is the plane, z is the axis and t (x_4) is the independent variable. In general, an axis of a **constant rotation** is a dimension that does not change under the rotation. Therefore, on SIX-SPACE an axis can have one, two or three dimensions; the plane of rotation can be of two, three or four dimensions; the independent variable can be of one, two or three dimensions.

In our proposed SIX-SPACE we have the alternative to consider the two hidden time dimensions as distinguishable or non-distinguishable. We adopt the second alternative. On Table 3.1 bellow we present all possible, distinguishable rotations on SIX-SPACE with two preferred dimensions. (For simplicity we use *x, y, z* for spacial coordinates and t, r, s for time coordinates.) The preferred space coordinate is *z,* and *t* the preferred time coordinate. Since they are preferred dimensions they are not naturally rotated, they are part of axis and/or independent variable. Planes are of 2, 3 or 4 dimensions; axes and independent variables are of 1, 2, or 3 dimensions. 32 distinguishable rotations are possible.

The configuration of plane, independent variable and axis is the fraction of Space (S) or Time (T) dimensions that they contain. Except for rotations 6 and 7, all rotations have different configurations. These configuration numbers are intended to serve as basis to identify the quantum numbers associated with each distinguishable rotation. For instance, a rotation with plane x,s is not different from one with plane y,s (both with plane configuration 1/2, 1/2) unless they differ in axis or independent variable configuration.

Line 1 of Table 3.1 represents a constant rotation whose axis of rotation is composed of three dimensions; the preferred space dimension z and the two hidden time dimensions r and s: The configuration is 1/3 space (S) and 2/3 time (T): The rotating plane is *x,y,* configuration is 2/2 =1 S, and 0/2 = 0 T; and the independent variable is the preferred time t, with configuration 0/1=0 S, and 1/1 = 1 T.

In line 2 the plane is x, y, the axis is t, r, s and z is the

independent variable. In both lines 1 and 2 the rotating

plane is pure space. In lines 3 and 4 the rotating plane is

pure time. In lines 5 and 6 the rotating plane is mixed 1,1

(one dimension of space and one of time). Preferred space

and time, z and t, exchange roles as axis and independent

variable. Table 3.1 shows that there are 32 distinguishable

modes of rotations on SIX-SPACE with two preferred

dimensions. We call these spinors.

Table 3.1 Distinguishable rotations on SIX SPACE

NUMBER	AXIS	CONFIGURATION		PLANE	CONFIGURATION		INDEPENDENT VARIABLE	CONFIGURATION	
		S	T		S	T		S	T
1	z, r, s	1/3	2/3	x, y	1	0	t	0	1
2	t, r, s	0	1	x, y	1	0	z	1	0
3	z, x, y	1	0	r, s	0	1	t	0	1
4	t, x, y	2/3	1/3	r, s	0	1	z	1	0
5	z, y, r	2/3	1/3	x, s	1/2	1/2	t	0	1
6	t, y, r	1/3	2/3	x, s	1/2	1/2	z	1	0
7	t, s, x	1/3	2/3	r, y	1/2	1/2	z	1	0
8	z, s, x	2/3	1/3	r, y	1/2	1/2	t	0	1
9	z, r	1/2	1/2	x, y	1	0	t, s	0	1
10	z, x	1	0	r, s	0	1	t, y	1/2	1/2
11	t, r	0	1	x, y	1	0	z, s	1/2	1/2
12	t, x	1/2	1/2	r, s	0	1	z, y	1	0
13	z, x	1	0	y, s	1/2	1/2	t, r	0	1
14	t, r	0	1	s, y	1/2	1/2	z, x	1	0
15	t, x	1/2	1/2	y, s	1/2	1/2	z, r	1/2	1/2
16	z, r	1/2	1/2	s, y	1/2	1/2	t, x	1/2	1/2
17	z	1	0	x, y	1	0	t, r, s	0	1
18	t	0	1	x, y	1	0	z, r, s	1/3	2/3
19	z	1	0	x, r	1/2	1/2	t, y, s	1/3	2/3
20	z	1	0	r, s	0	1	t, x, y	2/3	1/3
21	t	0	1	r, s	0	1	z, x, y	1	0
22	t	0	1	r, x	1/2	1/2	z, s, y	2/3	1/3
23	z, r	1/2	1/2	x, y, s	2/3	1/3	t	0	1
24	t, r	0	1	x, y, s	2/3	1/3	z	1	0
25	z, x	1	0	r, s, y	1/3	2/3	t	0	1
26	t, x	1/2	1/2	r, s, y	1/3	2/3	z	1	0
27	z	1	0	x, y, s	2/3	1/3	t, r	0	1
28	z	1	0	x, r, s	1/3	2/3	t, y	1/2	1/2
29	t	0	1	r, x, y	2/3	1/3	z, s	1/2	1/2
30	t	0	1	r, s, y	1/3	2/3	z, x	1	0
31	z	1	0	x, y, r, s	1/2	1/2	t	0	1
32	t	0	1	r, s, x, y	1/2	1/2	z	1	0

Rotation of the plane (x,y) is the negative of rotation of plane (y,x).This may be interpreted two different ways: 1- it determines the parity of a given spinor; or, 2- it is another spinor. If considered as a different spinor then the total number of distinguishable spinors would increase substantially.

Since in the theory that we are developing rotations on SIX-PACE and its dual are the constituents of all matter, then we conclude that there exist thirty (32) basic constituents of the physical world. (We are assuming that a change in parity does not lead to a different basic constituent). However, notice that rotation #3 of table 3.1 can be obtained from #2 by exchanging the roles of space and time coordinates. That is, #3 can be obtained from #2 by applying the Space-Time Operator, **K**, defined in equation (2.7). Symbolically, we may write 3 =**K** 2, meaning that rotation #3 on table 3.1 is obtained from #2 by applying the Space-Time operator **K**. Further analysis of table 3.1 shows that there are 16 such relations. They are tabulated on Table 3.4. For instance, rotation #4 = **K**#1.

<u>Table 3.4</u> First row is obtained from the second one by applying the Space-Time Operator.

#	4	3	7	8	12	11	14	16	21	20	22	26	25	30	29	32
K#	1	2	5	6	9	10	13	15	17	18	19	23	24	27	28	31

Thus, Table 3.1 can be divided into two tables, each one with 16 elements, and one is derivable from the other by applying the Space-Time Operator. On Table 3.5 we represent the 16 distinguishable rotations pertaining strictly to SIX-SPACE. Applying the Space-Time Operator to that table we obtain a second table with 16 elements pertaining to the dual space.

In present day theory of matter the basic constituents of matter are called **leptons, quarks and force carriers.** There exist 6 different **leptons, 6 quarks** and 4 **force carriers,** for a total of 16 basic constituents. Since we have obtained 16 basic spinors, we propose that these represent the six leptons, **electron, muon, taon** and their corresponding **neutrinos**; plus the six quarks **strangeness, up, down, charm, top, and bottom (s, u, d, c, t, b),** and the four force carriers known as **gluon, Z boson, W boson** and the **photon** of present day theory of matter. The photon belongs to SIX-SPACE and its dual and must have elements from both spaces. The rightness or wrongness of the theory can be tested by the possibility or impossibility to identify the spinors corresponding to each lepton and quark; also by the experimental proof that there exists (or does not exist) a cosmologically preferred space direction.

Anti-quarks and **anti-leptons** are the corresponding modes of rotations on the dual SPACE-TIME. Thus, we may say that the twelve dimensional SPACE-TIME that I am proposing consists of two sub-spaces: **SPACE** where matter exists, and **ANTI-SPACE** where anti-matter exists. Photons exist at the boundary of both sub-spaces, that is, they belong to both SPACE and ANTI-SPACE, and therefore the antiphoton does not exist. (Or, equivalently, a photon is its own antiphoton)

Table 3.5 Rotations strictly on SIX-SPACE with two preferred dimensions.

NUMBER	AXIS	CONFIGURATION		PLANE	CONFIGURATION		INDEPENDENT VARIABLE	CONFIGURATION	
		S	T		S	T		S	T
1	z, r, s	1/3	2/3	x, y	1	0	t	0	1
2	t, r, s	0	1	x, y	1	0	z	1	0
3	t, s, x	1/3	2/3	r, y	1/2	1/2	z	1	0
4	z, s, x	2/3	1/3	r, y	1/2	1/2	t	0	1
5	z, r	1/2	1/2	x, y	1	0	t	0	1
6	t, r	0	1	x, y	1	0	z, s	1/2	1/2
7	t, r	0	1	s, y	1/2	1/2	z, x	1	0
8	z, r	1/2	1/2	s, y	1/2	1/2	t, x	1/2	1/2
9	z	1	0	r, s	0	1	t, x, y	2/3	1/3
10	t	0	1	r, s	0	1	z, x, y	1	0
11	t	0	1	r, x	1/2	1/2	z, s, y	2/3	1/3
12	z, r	1/2	1/2	x, y, s	2/3	1/3	t	0	1
13	t, r	0	1	x, y, s	2/3	1/3	z	1	0
14	z	1	0	x, r, s	1/3	2/3	t, y	1/2	1/2
15	t	0	1	r, s, y	1/3	2/3	z, x	1	0
16	z	1	0	r, s, x, y	1/2	1/2	t	0	1

A most important step to follow from here is to establish a one to one correspondence among spinors with leptons, quarks and force carriers. The rotations corresponding to photons must have the peculiar characteristic of pertaining to both, SPACE and ANTI-SPACE.

In Figure 3.5 we represent the plane of rotations graphically, with circles. There are only five different kinds of plane configuration on Table 3.5: (1,0), (0,1), (1/2,1/2) (1/3, 2/3), (2/3, 1/3). They correspond to pure space, pure time; half time, halve space; 1/3 space, 2/3 time; 2/3 space, 1/3 time. They are enumerated in that order in Figure 3.5.

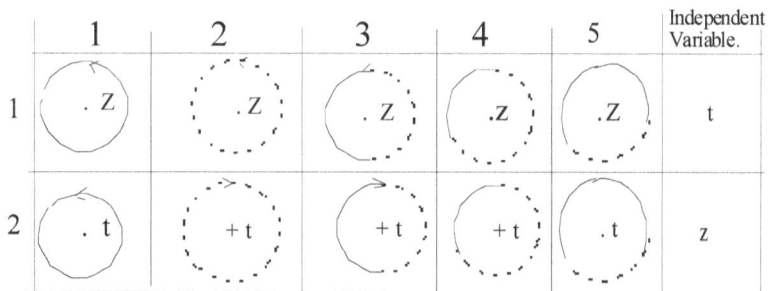

Figure 3.5 Graphical representation of rotations .

In rows 1 and 2 the roles of z and t are exchanged as axis and independent variable. Pure space plane rotations are represented, graphically, by solid circle; pure time plane by broken circle; mixed plane by part solid and part broken circle. A dot in the center indicates the axis of rotation pointing out of the page; a plus sign (+) represents an axis pointing toward the page. All rotations are drawn right handed; The independent variable and the axis must contain either t or z. For example, rotation (1, 1) of the figure is a pure space plane, t is in the independent variable and z is in the axis pointing out of the page (This represents rotations 1, and 5 on Table 3.5); (2, 1) is pure time plane, t is the independent variable and z is the axis pointing out of the page (represents rotation 9 of table 3.5); (3, 2) represents mixed 1,1 plane, t axis pointing toward the page, and z is the independent variable. In Table 3.6 all the rotations represented in Table 3.5 are identified with one of the graphical representations of Figure 3.5.

Table 3.6 **Correspondence of number of spinors on Table 3.5 to graphical representation coordinates on Figure 3.5**

Grahpical representation coordinates	1, 1	2, 1	3, 1	4, 1	5, 1
Spinors numbers	1, 5	9	4, 8, 16	14	12
Spinors numbers	2, 6	10	3, 11, 7	15	13
Graphical representation coordinates	1, 2	2, 2	3, 2	4, 2	5, 2

Notice the symmetry on the spinors distribution for each graphical representation coordinatess. This is a consequence of the total simetry of space and time in the theory.

3.3.2 Translation on six-space

We have postulated that all particles of nature are constituted of one or more rotations on SPACE-TIME. The group of rotations constituting a particle may displace on SIX-SPACE. Any infinitesimal displacement involves at least two dimensions and a derivative. For example, a Δx that occurs on a given Δt; and the component of velocity is the derivative of x respect to t. We now identify the different kinds of displacements on six-space.

1- Rest in SPACE and in TIME: Absolute rest. May not exist!

2-Rest in preferred TIME, uniform motion in SPACE: This represents electromagnetic waves, photons.

3-Rest in SPACE, uniform or nonuniform motion in TIME along preferred axis: (This includes a normal body at rest in space. Taking the big-bang as origin the existence of this state is questionable because all matter is moving in space relative to the big bang)

4-Rest in SPACE, uniform or nonuniform motion in TIME along hidden axis: Unknown.

5-Rest in TIME, uniform or nonuniform motion in SPACE: Unknown.

6-Uniform motion in SPACE and in TIME. Normal body in uniform motion.

7-Accelerated motion in SPACE, uniform motion in TIME: Includes orbital motion in space.

8-Accelerated motion in SPACE and in TIME: Possible orbital motion in both. Would imply the possibility of a return in space and time unless the orbit as such is moving.

3.4   The meaning of motion on six-space

In Minkowski space, where time has only one dimension, motion is represented by velocity and acceleration which are derivatives (changes) with respect to time. But in a theory that places time on equal footing as space one must take a more general approach. The position of any physical object or event will be represented by a point in SIX-SPACE. For example, a particle will be represented by a group of rotations centered at a given point in SIX-SPACE.

3.4.1 The position vector in SIX-SPACE

We start by recognizing that the coordinates (x_1, x_2, x_3, x_4, x_5, x_6) are just abstract, mathematical representation, or a set of independent variables for SIX-SPACE. The actual location of a particle or system is represented by a set of numbers $W = (w_1, w_2, w_3, w_4, w_5, \mathbf{w_6})$ in one to one correspondence to the coordinates (x_1, x_2, x_3, x_4, x_5, x_6). In general, each w_i is a function of the coordinates:

$$w_i = w_i\,(x_1,\ x_2, x_3,\ x_4,\ x_5,\ x_6) \text{ ---------------------- (3.6)}$$

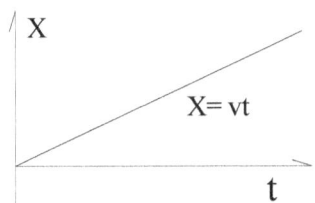

Figure 3.2 Particle path in
world-space.

As an illustration that the w_i are functions of the
independent variables $x_1...x_6$, let us consider a particle
moving with constant velocity, v, relative to the origin of
coordinates, along an x axis. The problem can be
represented on a two dimensional sub-space. (See Fig. 3.2).
In this case $W = (w_1, w_4) = (t, vt)$. Thus, both, w_1 and w_4 are
functions of the independent variable t. They do not depend
on x in this particular case.

From thermodynamics we can have another
illustration of the need to distinguish the independent
variables from the actual values that a system can have. The
natural thermodynamics space is represented by the
independent variables P, V, T for pressure, volume and
temperature, respectively. The state of any system is a point
in that space; $W = (w_p, w_v, w_t)$. For the particular case of an

ideal gas the equation of state is PV=nRT, and therefore w_p =nRT/V, w_v =nRT/P, w_t = PV/(nR). W is the vector in P, V, T-space whose components are the specific pressures, volumes and temperatures that the system can have. Any system different from an ideal gas will have a different set of possible values. Analogously, $(x_1, x_2, x_3, x_4, x_5, x_6)$ are the mere mathematical representation of SIX-SPACE, while $(w_1, w_2, w_3, w_4, w_5, w_6)$ are actual values assumed by the coordinates for a particular system in that space.

The physicist reader may convince himself that the above distinction is necessary by observing that in Minkowski-space the four coordinates, $x_1...x_4$, represent a set of independent variables. Yet, the coordinates of a physical event or system normally are functions of the fourth coordinate. By saying that the w_i are functions of some or all the independent variables, we are generalizing a fact known to be true in relativity for some of the coordinates. More important yet, is the fact that the position vector of a particle or system is one of the many vectors, like velocity, acceleration and force, whose components are functions of all or some of the independent variables.

The situation that we are presenting here is analogous to what we find in classical mechanics when we analyze a system of particles with constraints. For example, a system of two isolated particles that may be connected by a spring, a rigid rod, or by gravitational force, is a problem that involves six independent variables, three for each particle. Solving this problem, what is normally done is to say that the "position" of the system is a point on a six dimensional space. The interaction of the two particles -be it a connection with a rigid rod, a spring or a force field-gives us an equation from which we usually can express one of the coordinates in terms of the remaining five. By this action we simplify the problem reducing the six dimensional space by one dimension. That is, we reduce the number of independent variables. But we lose generality because the reduced space is not good for a system of two particles with different constraint. What we are saying is that we may keep the number of independent variables and use the constraints to find the components of the position vector as functions of the independent variables.

The fallowing question suggests itself naturally: Since Minkowski space is reduced from the more general SIX-SPACE that we are proposing, isn't it leaving out possible manifestations of the physical world that need five or six dimensions for their representation? Or else, hasn't the presumption that four-space is enough for all of physics lead to an erroneous conception of the physical world and of the cosmos? If we accept the premise that space and time are equivalent manifestations of SPACE-TIME, the answer to this questions should be a firm yes.

3.4.2 Generalized velocity on SIX-SPACE

The position point, W, is a tensor of the first rank, a vector, on six-space. We may write it as $W\alpha$. By covariant differentiation we obtain a second rank tensor $V\alpha\beta$,

$$V_{\alpha\beta}(x) = \frac{\partial W_\alpha}{\partial x_\beta} - [\alpha\beta, \gamma]W\gamma -------------(3.7)$$

where $[\alpha\beta, \gamma]$ is the Christoffell Symbol of the second kind (Repeated index implies summation). If the β index applies to the time coordinate of M_4, equation 3.7, applied to M_4, yields the three components of the velocity vector.

Therefore, $V_{\alpha\beta}(x)$ represents components of velocities in general. We define $V_{\alpha\beta}(x)$ as a **generalized velocity tensor**. Therefore, in SPACE-TIME, where time and space have equal standing, velocity is represented by a tensor with 36 components. (Most of these components may be zero). Given the position vector, to find the velocity of a particle or system in SIX-SPACE means to calculate the 36 components of generalized velocity using equation (3.7)

3.4.3 Representation of particles on SIX-SPACE

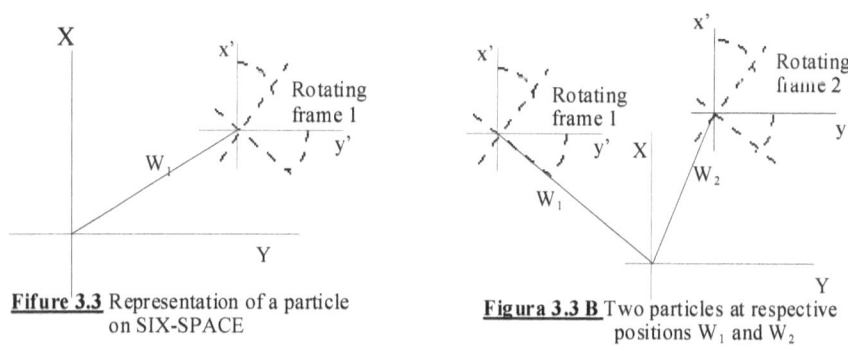

Fifure 3.3 Representation of a particle
on SIX-SPACE

Figura 3.3 B Two particles at respective
positions W_1 and W_2

On Figure 3.3 we represent a particle, **a group of SPACE-TIME rotations,** centered at some point in space. At that point coordinates X' are rotating and the origin of the rotating coordinates is the position, W, of the particle on the frame of reference X. Any other particle would have a similar representation. Two particles in the same space are represented in Figure 3.3-B. Interactions between particles separated in SPACE-TIME will be represented by the action that one rotation has upon another. For example, both particles may be located at the same time coordinates but at different space coordinates; or at the same space coordinate and at different time coordinates...The multiple possibilities for this situation should give us all the possible interactions -gravitational, electromagnetic, nuclear (strong or weak) - for the particles. Photons will be represented by rotations that are at rest on the preferred axes.

The different interactions will ensue from the different kinds of rotations and the separation, $W = W_1 - W_2$, of the particles. In the next chapter we present some speculative, graphical representations for the different interactions.

CHAPTER IV

FUNDAMENTAL FORCES GENERATED BY ROTATIONS

4.1   Rotations generating forces

On Figure 3.2 we saw that in an x, t plane a moving body has its position vector rotated from the t axis. Since an acceleration (a force) causes a change in velocity, it fallows that it causes a rotation. Conversely, we postulate, **rotations generate forces**.

We are talking about SPACE-TIME as such rotating. (Remember that SPACE-TIME is not a void). If two rotations coincide in SPACE-TIME they become a single rotation. Therefore, when we consider two particles there must exist a separation in their time and/or space location.

Any body, be it atom or planet, will have a pure time rotating plane, a pure space rotating plane and a mixed rotating plane resulting from the combination of all the rotations stemming from the spinors composing the particles. We postulate that these rotations are the causes of the fundamental forces of nature:

1-Pure time rotations located at the same time position but at different space positions cause gravitational force.

2-Combination of space and time rotations, at long distance in *space*, cause electrostatic forces.

3-Combination of space and time rotations, *at short distance in space*, cause strong interaction.

4-Pure space rotations at the same time position but with incoherent time rotations cause weak interaction.

All these will be illustrated graphically in the next section.

4.2   Graphical representation of interactions

In this section we represent, graphically, the mechanism by which fundamental forces are generated by rotations. In general, **attraction** will generate whenever two rotations reinforce each other; and **repulsion** when they go against each other.

4.2.1 Gravitational and electrostatic forces

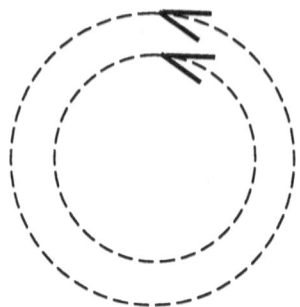

Figure 4.1 Two time rotations
at the same time position

Figure 4.1 represents two time rotations at the same time location. Their space location is different, say a distance R. This represents two masses at the same time coordinates but at different space coordinates. For instance, the small circle may be visualized as being further behind the page, at a distance R. The two time rotations, in the same sense, reinforce each other and are like a glue that links the masses. This is a graphical representation of postulate number one above. That is, gravitation is a time rotation phenomenon.

Electrical charges are the effect of time and space rotations. These rotations may be **coherent** (meaning that

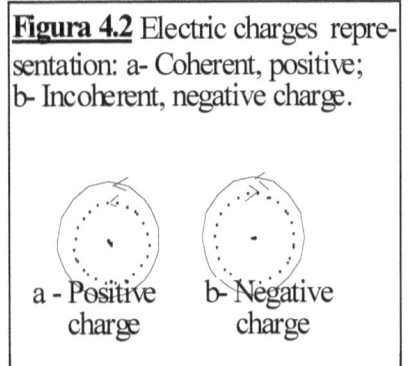

Figura 4.2 Electric charges representation: a- Coherent, positive; b- Incoherent, negative charge.

a - Positive charge

b- Negative charge

rotations are in the same sense) or **incoherent**: leading to **positive or negative** charge. Figure 4.2 represents this situation in which charges are separated in time and in space.

In Figure 4.3 we represent two elementary positive charges coincident in time, but at different space positions. Their respective centers of space rotations are separated a distance R. Notice that the two time rotations reinforce each other (Gravitational attraction), but the two space rotations are against each other. This leads to repulsion. The same happens with two negative charges.

Figura 4.3 Two positive electric charges .

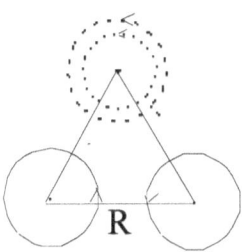

R

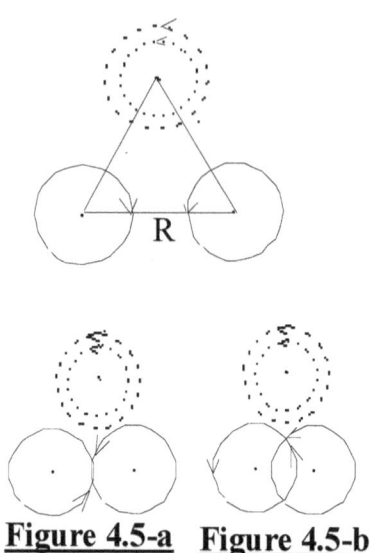

Figura 4.4 A positive and a negative electric charge .

Figure 4.5-a   Figure 4.5-b

Figure 4.4 represents a positive and a negative charge. Notice that in this case, as a result of the incoherency of the rotations of the negative charge, left circle in the figure, the two rotations now reinforce each other. This leads to attraction of opposite charges.

4.2.2 Nuclear forces (Strong and weak interaction)

In Figure 4.5-a we represent two positive charges coincident in time and very close in space. On this situation the opposition is maximum. But as they get closer in space, as shown in Figure 4.5-b their space rotations start reenforcing. The same rotations that were causing repulsion at long distance, at short distance cause attraction. This attraction cannot reach total coincidence in space, for then the two spinors become a single one. This suggests that strong interaction has the same origin as electrostatic interaction but acting at different space distances.

For the case of two opposite charges, say an electron and a proton, the interaction of the corresponding space rotations will be the same, but the time rotations will be opposite. This leads to a **weak interaction**, one that is stronger than the electrostatic interaction due to the short distance, but weaker than the strong interaction due to the opposite time rotations. The surprising result is that both, strong and weak interactions have the same origin as electrostatic interaction but operate at short distance.

CHAPTER V

FUNCTIONAL REPRESENTATION OF ROTATIONS ON SIX-SPACE

We have postulated that the interactions among rotations are the causes of the forces of nature. Specifically, the intersection of two rotations generate a force. In order to be able to find an equation representing a force generated by two specific rotations we must be able to represent each rotation in a mathematical function. The intersection of the two functions should lead to the corresponding equation. In this chapter we present a reasonable functional representation of rotations on SIX-SPACE.

5.1   Graphical representation of SIX-SPACE

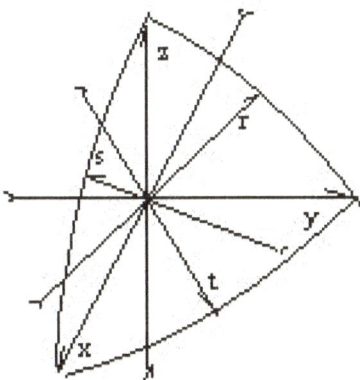

Figure 4.6 Graphical representation
of SIX-SPACE

Figure 4.6 is a graphical representation of SIX-SPACE. The three time axes are normal to the three space axes; For the sake of the representation they are drawn on perpendicular planes bisecting the three space planes. The functional representation of a rotation must contain an origin or center of rotation and an angular velocity distribution with symmetry about that origin. Fallowing a model of a rotation in a fluid, that velocity must start with a maximum value at the origin and decay to zero at a reasonable distance.

The functional representation must also conform to the different kinds of rotations represented on **Table 3.1 above.**

5.2   Proposed angular velocity distribution.

The functional representation must decay as one moves away from the origin along the rotating plane; It must also decay along the axis of rotation, but at a different rate. For example, if rotation is about the z axis and rotating planes parallel to the x,y plane, rotation #17 on Table 3.1, then there should be symmetry about the z axis and about the x,y plane. A function conforming these conditions is like a Fermi-Dirac distribution

$$F(x,y,z) = \frac{1}{e^{((\alpha z)^2 + x^2 + y^2)} + 1} -------------(5.1)$$

where α is a constant that allows the decay along the z axis to be different from that of the x,y plane. {Note: Any finite polynomial on x, y, z multiplying F (x,y,z) will also conform the conditions}. Setting the angular velocity at the source origin $= \omega_0$, we get

$$\bar{\omega}(x,y,z) = \frac{2\bar{\omega}_0}{e^{((\alpha z)^2 + x^2 + y^2)} + 1} \ \text{---} \ \text{---} \ \text{---} \ \text{---} \ \text{---}(5.2)$$

The spacial extension of the rotation, the boundary, would be the surface defined by the zeros of ω(x,y,z) on equation (5.2). This would mean letting the exponent go to infinity and the boundary is infinite, as is the case for the known fundamental forces of nature.

Equation (5.2) can be generalized for any axis, plane and independent variable contained in Table 3.1. This is achieved by substituting p_1 for the axis z, and p_2 for the plane. For example, for rotation 1 of Table 3.1 with axis z,r,s and plane x,y

$$p_1^{\ 2} = z^2 + r^2 + s^2 : p_2^{\ 2} = x^2 + y^2$$

and equation (5.2) becomes

$$\bar{\omega}(p_1, p_2) = \frac{2\bar{\omega}_0}{e^{((\alpha p_1)^2 + p_2^2)} + 1} \ \text{---} \ \text{---} \ \text{---} \ \text{---}(5.3)$$

The linear velocity distribution is the cross product

$$\vec{V} = \vec{\omega} X \vec{p}_2 = \frac{2\vec{\omega}_0 X \vec{p}_2}{e^{((\omega p_1)^2 + p_2^2)} + 1} \text{ } - - - - - - - - - - - - - - (5.4)$$

5.2 The interaction of two rotations

In section 4.2 we postulated that the interactions (the fundamental forces) are generated by the intersection of one rotation with another of the same kind. In this section we move one step forward suggesting the procedure that should be followed to obtain the laws governing those interactions.

Consider two rotations in SIX-SPACE representing two particles whose centers are separated by the spacial vector **R**, with

$$R^2 = x_0^2 + y_0^2 + z_0^2.$$

Set the origin of SIX-SPACE coordinates coincident with the center of rotation of one particle. If we wont to find their gravitational interaction, according to section 4.2.1 we must consider pure-time-rotation-planes. For example, rotation #3 on Table 3.1, which has (x,y,z) axis, (r,s) rotating plane and t independent variable. Let **V** be the lineal velocity distribution function for the rotation whose center is at the origin of coordinates; and **V'** that of the second rotation that is a distance **R**, along the axis, from the origin. Then, p_1 and p_2 for the corresponding velocity distributions on equation (5.4) are:

<u>For rotation at the origin</u>

<u>For the rotation at a distance **R**</u>

$$p_1^2 = x^2 + y^2 + z^2; \ p_2^2 = r^2 + s^2$$

$$(p_1')^2 = (x-x_0)^2 + (y-y_0)^2$$
$$+ (z-z_0)^2; \ (p_2')^2 = r^2 + s^2$$

We need to find **some kind of product** of the velocities distributions and to project them along **R.** If the integral over all volume of all those projection is in the direction of **R** that means repulsion; if opposite to the direction of **R** it means attraction. The force equation governing the interaction must derive from that integral. If the approach is correct, then it must give an inverse-square function of R for the law of gravitation. And a similar analysis for rotations representing electric charges must also lead to an inverse square law for relatively long distances.

As a first trial, one may take the cross product of the linear velocities **V** and **V'** for the product referred to in the above paragraph. Then the projection of the intersection of the two rotations on the vector **R** becomes

$$I = \oint \vec{V} \otimes \vec{V}' \bullet \frac{\vec{R}}{R} dv \; ------------------(5.5)$$

in which *dv* is a volume differential on SIX-SPACE. The challenge is to construct the products in the above integral on SIX-SPACE, and then perform the integration over all space.

According to the note fallowing equation (5.1), other reasonable trials would be with the velocities functions multiplied by reasonable, finite polynomials of R and/or p.

The challenge is placed. It is the turn for the daring physicists and/or mathematician, as mentioned in the PREFACE, to carry on.

Appendix I

Jesús Parrilla-Calderón

ROTATIONS AS GENERATORS OF FORCES

by

Jesús Parrilla-Calderón
Department of Physics
Cayey University College
Cayey, Puerto Rico.   00633

ABSTRACT

The situation of several rotating systems is discussed leading to the conclusion that rotations can be said to be the causes of forces in particular cases. Then we discuss the Lorentz transformation as a rotation and conclude that in general one can establish a one to one correspondence between any force and a rotation in a space where time coordinates can be rotated. Thus rotations may be said to be generators of forces.

Although some of the ideas are speculative in nature the author feels that they are sufficiently well founded on known physics and that they may stimulate other physicists to study the idea that rotations may be the generators of forces in nature.

Rotations as Generators of Forces

Introduction

One of the more striking demonstrations in mechanics to college physics students is the precession of a simple bycicle wheel. Although this is nothing but a sofisticated top the student can never anticipate what will happen to the wheel when its axis is held by one of its ends in such a way that it can rotate in a horizontal plane as well as in a vertical plane. See figure 1.

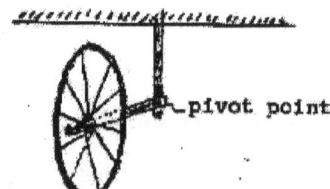

Figure 1. Wheel can rotate about the principal axis and about the pivot point.

When one holds the axis horizontally and releases it suddenly the wheel simply falls down, its axis rotates until it assumes a vertical position as shown in figure 2.

Figure 2. No rotation, wheel falls down.

-2-

But if the wheeel is given a rotation about its axis before it is released, then it does not fall but rather precesses on a nearly horizontal plane. It is not surprising to hear the exclamation "Rotation of the wheel about its axis has produced a force that prevents it from falling".

In this article we analyze the situation of various rotating bodies and conclude that indeed one can explain the situation by attributing to rotations the production of forces. The question as to whether this is a mere representation with no possible physical consequence is analyzed in the case of solids rotating with a liquid and an experiment is suggested where the absence of presence of physical consequences may be established. We end by using the Lorentz transformation to generalize the situation associating a rotation to any force.

Force generated by one rotation around one axis

The case of the centrifugal force due to a rotation of a body in space stands as a witness to the fact that rotations can cause forces. The equation for centrifugal force

$$\vec{F}_c = m\omega^2 \vec{r} \qquad (1)$$

means that whenever a body of mass m rotates with angular frequency ω on a circle of radius r, seen from its own frame of reference it experiences a force $\vec{F}_c$. One

73

usually states that this is only a fictitious force, that
the real one is the centripetal force which prevents the
body from following a straight line. This is true from
the view point of the   nonrotating system of reference.
But from a rotating system point of view both forces are
real. This is clearly illustrated by the situation of
an astronaut in a closed cylindrical capsule which is
rotating around its axis. If the astronaut ignores the
fact that the capsule is rotating then the only thing he
knows is that in his spaceship there exist a force field
that pulls him and every object toward the inside of the external
wall of the capsule . If the radius of the capsule is
sufficiently large, so that for the astronaut the surface
is flat, then what the astronaut observes is that all
bodies in his space are pulled to the wall with a force
which is proportional to the mass of the body. He may
develop proyectile kinematics identical to the one developed
here on Earth. He may find some anomalies of the field
as he surveys his space along different orientations, but
he can attribute these anomalies to "slight" curvature of
his space along some directions. As long as he cannot
notice a curvature of his space he will not notice anomalies,
for his space will be an infinitesimal segment of the
cylinder surface and this is almost a point on the surface.

   We, observers of the capsule from a   nonrotating

system, say that the astronaut is subject to a centripetal force and that the centrifugal force is a fictitious one. But the astronaut and all bodies in the capsule feel the effect of that force we call fictitious just as well as we feel the effects of gravity on Earth. Thus, because a body is rotating relative to the rest of the universe, with constant angular velocity and radius; we non-rotating observers conclude that it is experiencing a centrifugal force, very real to him, given by equation (1). That equation shows that the force is indeed proportional to the mass, the proportionality constant being $\omega^2 r$.

Whether a force is real or not should be decided on the basis of the effects felt by bodies in which the force acts and not on how the force is percieved by outside observers. Thus we shall conclude that centrifugal forces are physical. In a sense then, they are as fundamental as gravitational forces. In fact it is well known that centrifugal forces are attributed to the effect of very distant masses upon bodies. (Mach's principle)

Force generated by two rotations around perpendicular axes

We shall now consider the precession of the bicycle wheel of figure 1. The Euler angles θ , ϕ , ψ specify the orientation of the symmetry axis as shown in figure 3. It is well known that as seen by a non-rotating

-5-

frame of reference the vertical motion of the loose end

of the axis is a nutation(¹) (²). In the particular

case in which the body is simply released

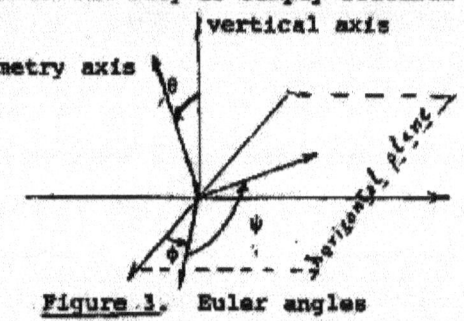

symmetry axis

Figure 3. Euler angles

with no initial horizontal velocity (The only inital

motion is rotation around the symmetry axis) the nutation

will describe a curve called cuspidal curve shown in

figure 4. The axis nutates between the limiting angles

θ_1 , and θ_2 . If the initial conditions are changed

the type of nutation also changes. The critical situation

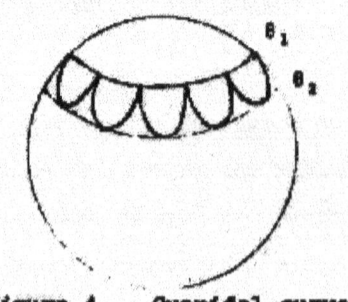

Figure 4. Cuspidal curve.

will be the one with no nutations. This means $\theta = 0$
and $\theta_1 = \theta_2 = \theta_c$. We shall return to this particular
case latter on.

One of the cups of the cuspidal motion is represented
in figure 5 as a series of straight segments. Any
individual segment may be said to represent the combination
of two rotations on perpendicular axes. For example, for

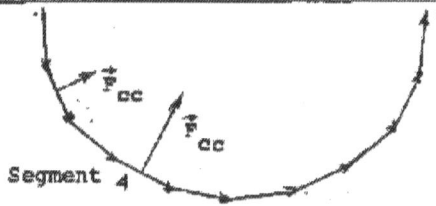

Figure 5

A single cup of a cuspidal curve represented in

straight segments.

segment 4 on figure 5 the body is rotating instantaneously
around the symmetry axis (primary axis) which points out
of the page, and through a secondary axis perpendicular
to the symmetry axis, passing through the point of rotation
or origin, and perpendicular to the segment. The motion,
as seen from a none rotating frame, is suggestive of a
motion under a variable central force as $\vec{F}_{cc}$ in figure 5.
Therefore the motion may be represented by that due to
a force, $\vec{F}_{cc}$, generated by the combined effect of the
two rotations pointing along the instanteous direction

of the secundary axis of rotation. ($\vec{F}_{cc}$ for double rotation, $\vec{F}_c$ for single rotation).

The problem of finding the force which is a function of the two rotations alone is simplified by taking the critical case where $\theta = \theta_c = \pi/2$ and the precession is without nutations. In that case the force, $\vec{F}_{cc}$, does not change direction, pointing vertically upwards.

Let I_1 and I_2 be the moment of inertia about the symmetry axis and about the vertical axis respectively. Then the condition that at $t = 0$ both $\dot{\theta}$ and $\ddot{\theta}$ shall be zero is [3]

$$mg\ell = \dot{\phi}(\dot{\psi}I_1 - (I_2 - I_1)\dot{\phi}\cos\theta) \qquad (2)$$

where ℓ is the distance from the fixed point (origin) to the center of mass. If in addition we wish to have $\theta = \theta_c$ then (2) is valid at all time, and for $\theta_c = \pi/2$ it becomes

$$mg\ell = \dot{\phi}\dot{\psi}I_1 \quad . \qquad (3)$$

As seen from a system which is rotating with the wheel around the vertical axis a force diagram is as shown in figure 6.

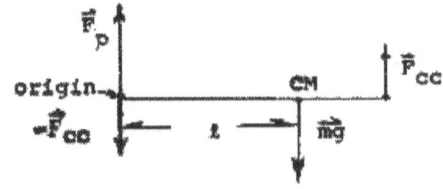

Figure 6
From the rotating frame F_{cc} is real.

−8−

The law of conservation of momentum demands that F_{cc} act also on the pivot point in the opposite direction. F_p is the vertical component of the force made by the pivot point. F_{cc} must be real because the system does not rotate about the origin under the action of the weight mg. Thus the torque mgℓ is balanced by the torque F_{cc} k where k is the distance from the point of rotation to the point of action of $\vec{F}_{cc}$. A reasonable assumption would be that k is the radius of gyration of the body (the wheel) about any axis passing through the origin perpendicular to the symmetry axis. Then k is given by k = $(I_2 /m)^{1/2}$. The equilibrium equation leads to

$$mg\ell = F_{cc}k \qquad (4)$$

and using (3) and the value of k we obtain

$$F_{cc} = \dot{\phi} \; I_1 \; \dot{\psi}(m/I_2)^{1/2} \qquad (5)$$

Since the direction of $\vec{F}_{cc}$ is along $\dot{\phi}$ we can write it as

$$\vec{F}_{cc} = F_{cc} \; \frac{\vec{\phi}}{\phi} \qquad (6)$$

Substituing (5) on (6) we finally obtain

$$\vec{F}_{cc} = (I_1^2 m/I_2)^{1/2} \; \dot{\psi} \; \vec{\dot{\phi}} \qquad (7)$$

Notice that $\vec{F}_{cc}$ is sensible to changes in sign of $\dot{\psi}$. Thus we assume that it is positive (negative) when $\vec{\dot{\phi}}$ points away from (towards) the origin.

Equation (7) shows that for this particular case one can conclude that the two combined perpendicular rotations generate a force on the body. Like the centrifugal force, this one looks like a fictitious force for non-moving observers, but, for the body that feels it, it is real and if its torque is not balanced by some other force, like gravitation in this case, the body will rotate obeying the force $\vec{F}_{cc}$.

Equation (7) is not symmetric with respect to exhange of the two rotations. But this is to be expected since one can distinguish one rotation from the other, since $\dot{\psi}$ is the rotation about the symmetry axis. (Also $\dot{\psi}\,\vec{\dot{\psi}}$ is a dot product of the unit vector $\dot{\psi}/\psi$ and the dyad $\vec{\dot{\psi}}\,\vec{\dot{\phi}}$, and this does not commute.)

We can generalize equation (7) to any two rotations one which is about the principal axis, $\vec{\alpha}_1$, and a secundary one, $\vec{\alpha}_1$, perpendicular to the first, with respective moments of inertia I_1 and I_2 . The force $\vec{F}_{cc}$ will be given by

$$\vec{F}_{cc}\,(\dot{\alpha}_1,\,\vec{\dot{\alpha}}_2) \;=\; (I_1^2 m/I_2)^{1/2}.\; \dot{\alpha}_1\,\vec{\dot{\alpha}}_2 \qquad (8)$$

Equation (8) can be corroborated by using two
masses with a common symmetry axis rotating in opposite
directions as represented in figure 7. The symmetry
axis can rotate upward on the pivot point but the arms
prevent them from doing this. When the force F_{cc} tends to

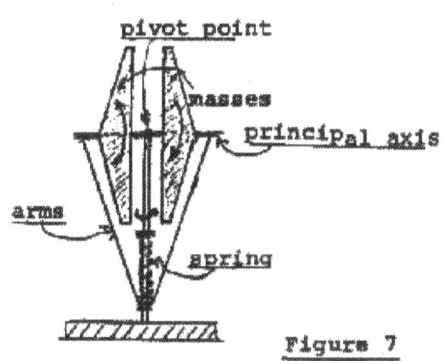

Figure 7

F_{cc} pulls the masses up and can
be measured on the spring

move the masses up the spring compresses and F_{cc} can be
measured.. The author has used a system like this to
detect F_{cc}.

The case of a general nutation can now be analized
by using (8) applied to infinitesimal segments of the curve.
In any segment the conditions needed for the applicability

-11-

of (8) are satisfied instantaneousiy. That is, in any
segment we can identify two perpendicular rotations, $\vec{\alpha}_1$.
and $\vec{\alpha}_2$, and corresponding constant moments I_1 and I_2,
α_1 is the Euler angle ϕ but α_2 is not ϕ. But since we
are dealing with infinitesimal segments we can deal with
α_2, ϕ and θ as vectors, as shown in figure 8. Part

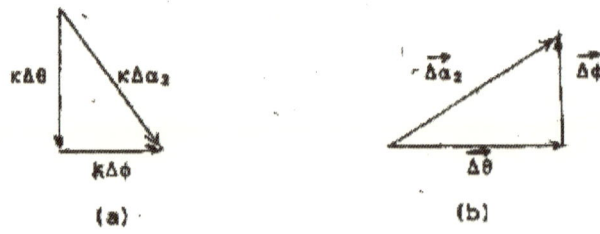

(a)   (b)

Figure 8

(a)   Segments $k\Delta\alpha_2$ formed by segments $k\Delta\theta$
      and $k\Delta\phi$.

(b)   Corresponding angles triangle .

(a) shows the segment $k\Delta\alpha_2$, formed by the rotation at the
radius k, composed of the two perpendicular segments $k\Delta\theta$
and $k\Delta\phi$. The corresponding relationship for the infiniti-
tesimal vector angles is shown in (b). We see that
$\vec{\Delta\alpha}_2 = \vec{\Delta\theta} + \vec{\Delta\phi}$ from which we obtain

$$\vec{\alpha}_2 = \vec{\theta} + \vec{\phi} \qquad \alpha_2 = \left(\dot{\theta}^2 + \dot{\phi}^2\right)^{1/2}$$

Substituing on (8) we obtain

$$\vec{F}_{cc}\ (\ \dot{\vec{\psi}},\ \dot{\vec{\theta}},\ \dot{\vec{\phi}}\ )\ =\ (\ I^2_1\ m/I_2\ )^{1/2}\dot{\vec{\psi}}\ (\dot{\vec{\theta}}\ +\ \dot{\vec{\phi}}).\quad (9)$$

Equation (8) is a force generated by a rotation in
two angular dimensions and (9) is for three angular
dimensions. (9) is symmetric with respect to exchange
of the two secondary rotations $\dot{\theta}$ and $\dot{\phi}$, but not with
respect to the primary rotation $\dot{\psi}$, and the other two.

Force generated by two rotations on one axis, or differences in rotations.

The case of tea leaves in a tea pot which is known
to have captured Einstein's attention, and the formation
of meanders by rivers are cases where difference in
frequency of rotation around the same axis causes a
force that pushes the slower part toward the center of
rotation, against the centrifugal force.

Solid grains, or pieces suspended in a liquid,
rotate slower than the liquid when one sets the liquid
into rotation. This happens because what sets the suspended
solids into rotation is the flow of the rotating fluid
and this can never give to the solids a frequency higher
than that of the liquid. The easily observable result
is that the solids move toward the axis of rotation,
going against the centrifugal force.

From the frame of reference of the suspended solids
what happens is that the fluid is rotating, and that causes

on them an inward force. From the non-rotating frame
of reference the force is caused by the difference in
rotation. Thus we can talk of a third force $\vec{F}_d$ which is
caused by rotations, in this case a difference in rotations.

The general case of a fluid spinning around one axis,
has been studied by many. [*] If a layer of the fluid
spins slower than the rest of it then the slower layer
of fluid moves toward the axis of rotation establishing
complicated secondary flows until the system reaches
a new equilibrium. Again, from the view point of the
system at rest with the slower fluid, a real force
develops as a result of the rotation of the rest of the
fluid. From the stationary system one sees that that
force is due to the difference in rotation of the two
layers of fluid.

The forces $\vec{F}_d$ acting on suspended solid particles
as mentioned above are said to be manifestations of
viscous forces that secondary-inward-fluid-flows makes
on the solids. [*] Since the secondary flux is caused
by a difference in frequency of layers of fluid then,
in principle, one can establish a one to one correspondence
between $\vec{F}_d$ and a difference in frequency. The question
is, which difference in frequency: Is it between the
liquid and solids or is it between two layers of liquid?
If it happens to be between liquid and solids then the

-14-

force $\vec{F}_d$ is as real as $\vec{F}_c$. Well controlled experiments
with solids suspended in rotating liquids could then
establish if the force $\vec{F}_d$ does or doesn't have physical
reality.

General forces and rotations

The Lorentz transformation represents a rotation

$$X' = L X \tag{10}$$

in Minkowski space. The angle of rotation, θ, is
given by $\cos\theta = \gamma$; $\text{sen } \theta = i\beta\gamma$; $\gamma = (1 - v^2/c^2)^{-1/2}$,
$\beta = v/c$. Allowing for small variations of v so that
(10) still be valid, we can differentiate with respect
to the stationary-system-time-coordinate obtaining

$$d \cos\theta/dt = d\gamma /dt = \gamma^3 (v/c^2)dv/dt,$$

$$-\text{sen}\theta \, d\theta/dt = (\gamma^3 v /c^2) \, dv/dt$$

$$d\theta/dt = (i\gamma^2/c) \, (dv/dt)$$

from which we obtain

$$d\theta/dt = (i\gamma^2/c) \, (dv/dt)$$

or

$$\dot{\theta} = (i\gamma^2/c)\, a. \tag{11}$$

Equation (11) shows that whenever the "stationary"
observer sees an accelerated body he can establish a
one to one correspondence between the acceleration and a
rotation. A strange rotation however, since it involves

85

the rotation of a time coordinate. Conversely, any
rotation of the space-time coordinates reflects an
acceleration of bodies and this in turn shows the
presence of forces.

If one does not want to use Minkowski space ("Good-
by to it").[*] the Lorentz transformations pass from a
rectangular frame into an oblique one. This represents
a rotation of two axes toward each other, but still a
rotation, and a one to one correspondence like that of
equation (11) can still be established.

From the above, paragraphs it is evident that
if we work on a physical space where time coordinates
can be rotated, then we can associate a physical rotation
to any acceleration and therefore to any force. The
question then arises, shouldn't it be possible to
develop a relativistic theory of matter where all forces
are represented by suitable rotations on a more general
space where time rotations can occur? Such a theory of
physics will then change the equations of dynamics,
which are said to consist of one side of kinematical
terms equated to a side containing the causes of motion
or forces.[*] They would be changed into all
kinematical terms with the forces substituted by rotations
on a generalized space. This may be what unifying field
theories are pointing out to physicists, since all these

-16-

theories involve spinors (rotors) on general spaces of
(2) (4) (9) (16)
unknown physical nature.

Jesús Parrilla-Calderón

1. H. Goldstein, <u>Classical Dynamics</u>, Addison Wesley (1950).

2. W. Arthur and S. K. Fenster, <u>Mechanics</u>, Holt Rinehart and Winston (1969).

3. H. Goldstein equation (5-70).

4. See for example J. D. Baker, Jr.; <u>Am. J. Phys. 36</u>, 980 (1968) and references thereby.

5. C. W. Misner, K. S. Thorne and J. A. Wheeler, <u>Gravitation</u>, W. H. Freeman and Company, San Francisco (1970).

6. For example, E. C. G. Sudarshan, personal communication.

7. H. Georgi, S. L. Glashow, <u>Phys. Rev. Lett. 32</u>, 438, (1974).

8. S. Weinberg, <u>Phys. Rev. Lett. 42</u>, 850 (1979).

9. K. Hayaski, T. Shirafuyi, <u>Phys. Rev. D 19 No. 12</u>, 3524 (1979).

10. W. Mecklenburg, <u>I. C. T. P.</u>, IC/79/87 and references thereby.

88

AMERICAN
JOURNAL
of PHYSICS

A Journal of the American Association of Physics Teachers

John S. Rigden, Editor
Philip B. James, Assistant Editor

Room 240 Benton Hall
University of Missouri - St. Louis
St. Louis, Missouri 63121 U.S.A

July 30, 1980

Professor J. Parrilla-Calderon
Department of Physics
Catholic University of Puerto Rico
Ponce, PUERTO RICO 00731

Dear Professor Parrilla-Calderon:

Two reviews of your manuscript "Rotations as Generators of Forces" are enclosed. It is quite clear from these reviews that your work is too speculative for publication in a pedagogical journal such as the American Journal of Physics. Therefore, we shall not be able to accept this manuscript for publication.

Sincerely,

Philip B. James
Assistant Editor

CC: Referees

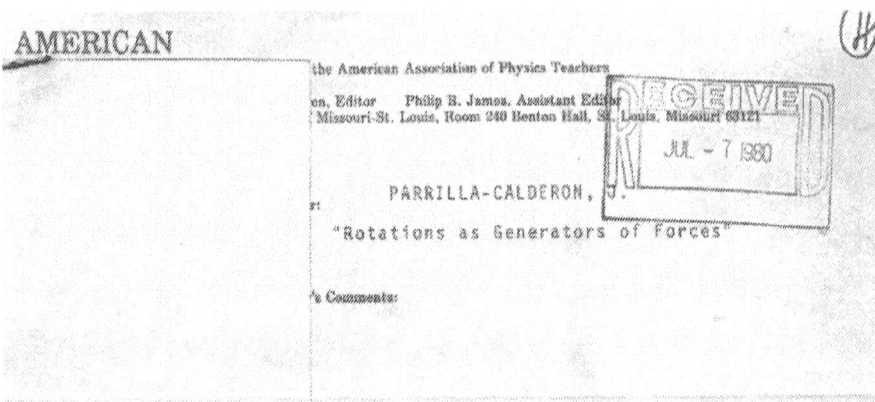

the American Association of Physics Teachers

on, Editor   Philip B. James, Assistant Edi
Missouri-St. Louis, Room 240 Benton Hall, St. Louis, Missouri 63121

RECEIVED
JUL - 7 1980

PARRILLA-CALDERON, J.

"Rotations as Generators of Forces"

's Comments:

... for common manuscript ... been submitted to the American Journal of Physics. The Editors will appreciate your opinion regarding its suitability for publication in the Journal. Detailed comments may be included on the reverse of this sheet.

Since no further action can be taken on this paper until your review is received, your attention to this matter as soon as possible will be appreciated. Please return this form, along with any manuscript pages you have commented upon, in the enclosed envelope. Keep the manuscript on file for at least six months. Remember, the manuscript should be regarded as confidential.

1. Briefly summarize this manuscript for the editors emphasizing why its contributions would be of interest to the AJP readers.

This paper tries to present an alternative picture of gyroscopic motion. However, it falls far short of its aim. The first four pages are a more or less correct low-level discussion of centrifugal force. Pages 5 to 12 try to generalize this concept to gyroscopic motion. Some comments on this are: i) By Fig. 6 force equilibrium implies $k = \ell$, not the radius of gyration; ii) by the same considerations $F_{cc} = mg$, so why does the author express F_{cc} in terms of $\dot{\phi}$ and $\dot{\psi}$ as in Eq. (7)? An explicit evaluation of Eq. (7) will give $F_{cc} = mg$ again. On page 12 the author explains the tea cup effect in terms of the tea leaves lagging the fluid, but then on the next page he quotes a different (and more correct explanation). The last
(over)

2. Is the manuscript technically correct?

On page 10 an apparatus is briefly discussed in which F_{cc} can be measured. If the author has (as he claims) detected F_{cc} to be non-zero, it would be a great boon to mankind. This measurement of levitation, or its expectation, demonstrates the incorrectness of his argument.

3. Are the style, grammar, level, etc. of this manuscript suitable for publication?

OK, except for numerous spelling errors.

4. Overall recommendation
_____ Enthusiastically recommend publication
_____ Recommend publication

_____ Can be published with revision
        _____ I should see the revision
        _____ I don't need to see the revision
__X__ Reject the manuscript

section (pages 14 to 16) is even more speculative and vague, especially at the end. This paper is partly pedantic, but mostly presents the author's unconventional ideas and as such is not suited for the AJP.

Jesús Parrilla-Calderón

Appendix II

Jesús Parrilla-Calderón

nature

Macmillan Journals Ltd
4 Little Essex Street
London WC2R 3LF
Telephone 01-836 8633
Telex 262024

Dr. J. Parrilla Calderon
Department of Physics
Cayey University College
Cayey PR 00633
USA

Reference No 11019   06 Nov 80

Dear Dr Parrilla Calderon,

Thank you for the manuscript you have submitted to Nature.
We will be in touch with you as soon as we are able to
reach a decision on it. In the event of any query please
quote the above reference number.

Editorial Office
Nature

95

Jesús Parrilla-Calderón

nature

Macmillan Journals Ltd
4 Little Essex Street
London WC2R 3LF
Telephone 01-836 6633
Telex 262024
Advertising 01-240 2044

In reply please quote:
11019   KG/SC

2 December 1980

Dr. J. P. Calderon
Departamento de Matematica Fisica
Universidad de Puerto Rico
Colegio Universitario de Cayey
Cayey Puerto Rico 00633

Dear Dr. Calderon,

Thank you for submitting your manuscript "Force Generated by Two Perpendicular Rotations II" for consideration. Although we had no specific
scientific criticism of your paper, we were of the firm opinion that your
paper was of insufficient immediate interest to demand publication in
Nature. With regret I am therefore returning your manuscript so that
you can submit it to a more appropriate journal such as the American
Journal of Physics.

The competition for our limited space is, as always, very great and it is
inevitable that we can only select those papers that seem to offer the
most to our readers.

I am sorry that we cannot be more positive.

Yours sincerely,

Dr. Konrad Guettler
Physical Sciences Editor.

Enc.

96

FORCE GENERATED BY TWO PERPENDICULAR ROTATIONS II

BY

Jesús Parrilla-Calderón
Department of Physics
Cayey University College
Cayey, Puerto Rico   00633

Abstract

In this article, following a previous article titled Force Generated by two Perpendicular Rotations [1] we show theoretically that two perpendicular rotations must generate a force. Then we present pictures of an apparatus which we built in order to show experimentally the presence of that force. In a sequence of pictures of the apparatus we show this force in action and demonstrate the way in which this force can be measured. The results may find applications in physics and technology since extremely large forces can be generated with suitable rotations.

-2-

Theory

In this article we show that if a rigid body which has a symmetry axis has two simultaneous perpendicular rotations , with one point fixed, then a fictitious force, $\vec{F}_{cc}$, due to this rotations alone , and aside from the centrifugal force $\vec{F}_c$, must act on the body. To do this we shall use the fundamental relationship between time derivatives of a vector for rotating systems

$$(d\vec{A}/dt)_s = (d\vec{A}/dt)_r + \vec{\omega} \times \vec{A} \qquad (1)$$

where $(d\vec{A}/dt)_s$ and $(d\vec{A}/dt)_r$ stand for the time-rate-of change of $\vec{A}$ as seen from the stationary and rotating system respectively. [2] $\vec{\omega}$ is the angular velocity of the rotating system.

Consider a body rotating with constant angular velocity ω_1 about its symmetry axis, and ω_2 about an axis perpendicular to the symmetry axis. As seen from the stationary system it has angular momentum

$$\vec{L} = I_1 \vec{\omega}_1 + I_2 \vec{\omega}_2 \qquad (2)$$

where I_1 and I_2 are the moments of inertia corresponding to the axis of $\vec{\omega}_1$ and $\vec{\omega}_2$. The products of inertia are zero due to the symmetry of the body. Since $\vec{\omega}_1$ is changing direction then $\vec{L}$ is varying with time. Then equation (1) applies to this case where $\vec{\omega} = \vec{\omega}_2$ and $\vec{A} = \vec{L}_1 = I_1\vec{\omega}_1$. Thus

-3-

$$(d\vec{L}_1/dt)_s = (d\vec{L}_1/dt)_r + \vec{\omega}_2 \times \vec{L}_1 \tag{3}$$

From the rotating frame of reference the body rotates only about the symmetry axis with constant $\vec{\omega}_1$ and therefore $(d\vec{L}/dt)_r = 0$. Thus

$$(d\vec{L}_1/dt)_s = \vec{\omega}_2 \times \vec{L}_1 = \vec{\omega}_2 \times (I_1\vec{\omega}_1)$$

From Newton's second law applied to rotations the above equation becomes,

$$\vec{\zeta}_c = \vec{\omega}_2 \times \vec{\omega}_1 I_1 . \tag{4}$$

Equation (4) means that for a body to have two perpendicular rotations, $\vec{\omega}_1$ and $\vec{\omega}_2$, a torque $\vec{\zeta}_c$ must act on it. This is analogous to the situation found for a particle in uniform circular motion. In that case one finds that for the particle to have that motion a centripetal force $\vec{F}$ must act on it. This leads to the conclusion that from the point of view of a system rotating with the particle there is a centrifugal force $\vec{F}_c = -\vec{F}$ acting on the body. By analogy, from the frame of reference rotating with velocity $\vec{\omega}_2$ a torque $\vec{\zeta}_{cc} = -\vec{\zeta}_c$ must act on a body that has two simultaneous and perpendicular rotations $\vec{\omega}_1$ and $\vec{\omega}_2$.

Any torque may be said to be produced by a cuople, that is , by a force which acts at two different points in opposite directions. This means that the torque $\vec{\zeta}_{cc}$

-4-

must be due to a force F_{cc} which acts at two different pointsassociated to the body. The motion is with one point fixed. This is a very important point where forces of constraint must be acting. Equally important is the radius of gyration measured from the secondary axis. It is logical to assume that the force F_{cc} acts on these two points, perpendicular to the symmetry axis and on opposite directions. Then the net force is zero as demanded by the law of conservation of lineal momentum, but the torque is not. The torque will then be

$$\vec{\zeta}_{cc} = \vec{\kappa} \times \vec{F}_{cc} \tag{5}$$

where $\vec{\kappa}$ is the radius of gyration.

We wish to obtain an expression for $\vec{F}_{cc}$ by using (4) and (5). To do this cross multiply (5) by $\vec{\kappa}$ obtaining

$$\vec{\kappa} \times \vec{\zeta}_{cc} = \vec{\kappa} \times (\vec{\kappa} \times \vec{F}_{cc}) = \vec{\kappa}(\vec{\kappa} \cdot \vec{F}_{cc}) - \vec{F}_{cc}\kappa^2$$

Since $\vec{\kappa}$ is perpendicular to $\vec{F}_{cc}$ we obtain

$$\vec{F}_{cc} = -\vec{\kappa} \times \vec{\zeta}_{cc}/\kappa^2 \tag{6}$$

Using (4) and $\vec{\zeta}_{cc} = -\vec{\zeta}_{c}$ we obtain

$$\vec{F}_{cc} = \vec{\kappa} \times (\vec{\omega}_2 \times \vec{\omega}_1) I_1/\kappa^2$$

which expanding and considering that $\vec{\kappa}$ is perpendicular to $\vec{\omega}_2$ and that $m\kappa^2 = I_2$, leads to

$$\vec{F}_{cc} = (I_1^2 m/I_2)^{1/2} \omega_1 \vec{\omega}_2 \tag{7}$$

100

-5-

Equation (7) means that whenever a symmetric body rotates about its symmetry axis with angular velocity ω_1 and simultaneously rotates with angular velocity $\vec{\omega}_2$ about an axis perpendicular to the symmetry axis, keeping one point fixed, then a force $\vec{F}_{cc}$ acts on the body on the point of gyration, directed along the secondary axis .

Equation (7) can be generalyzed to cover any case in which $\vec{\omega}_1$ has a component perpendicular to $\vec{\omega}_2$. Then in (7) one substitutes the perpendicular component of $\vec{\omega}_1$ instead of ω_1 .[3]

Experimental verification

The force F_{cc} has been detected by means of the apparatus shown in picture-1. The two bicycle wheels shown on the picture are mounted on nearly horizontal axis. Their axes can rotate upwards but two adjustable metal rods prevent them from moving up unless the force is strong enough to compress the spring. The spring constant is about 200 lbs./in. , and the spring can compress for about one inch. The wheels and spring are mounted on a Sargent Welch # o930 centripetal-force-apparatus permitting simultaneous rotations about the wheels axes and about the centripetal-force-apparatus axis (Secondary axis). From equation (7) these simultaneous rotations must produce a force, $\vec{F}_{cc}$, which will tend to rotate the wheel axes upward and the spring will compress.

101

-6-

Picture-2 is a close-up on the spring to show also a white tape covering the axis inmediately below the spring. This serves as a clearly visible reference point to observe when the spring compresses while the system is moving. Picture-3 shows the two wheels rotating about their principal axes. This is evident from the fact that the rays cannot be seen . The wheels are rotating in opposite sence at nearly equal frequencies so that the total angular momentum of the system is zero. Picture-4 shows the wheels rotating about the secondary axis but with no rotation about the prrincipal axis. On pictures-3 and 4 the white tape is right below the spring showing that it has not compressed in spite of the separated rotations.

Picture-5 shows the wheels rotating about their principal axes while motion about the secondary axis is building up. Motion is given to the system by hand. Notice the wheels pulling up and the spring compressed as evidenced by the metalic luster sowing between the lower part of the spring and the white tape. Picture-6 shows the system at considerable speed about both axes and the spring compressed nearly one inch as shown by the metalic luster between the spring and the white tape. (Compare with picture-2). This means that the wheels are pulling the spring with about one hundred pounds each. The weight of each wheel is six pounds, thus the force has nothing to do with gravity. In fact, changing the orientation of the secondary axis , setting it vertically down or horizontal , has negligible efect on the spring. Picture-7 shows the system with most of its motion lost due to friction and the spring returning to its original length but still a little compressed by the remnant motion.

REFERENCES

1- J. Parrilla-Calderón , article submitted to Foundations of Physics .

2- H. Goldstein, Classical Mechanics, Adison-Wesley (1950) equation (4-100) ; W. Arthur and S. K. Fenster, Mechanics, Holt Rinehart and Winston,(1969) equation (2.92) T. R. Kane, Dynamics, Holt Rinehart and Winston, (1968) equation 2.13 .

3- J. Parrilla-Calderón , op. cit.

Jesús Parrilla-Calderón

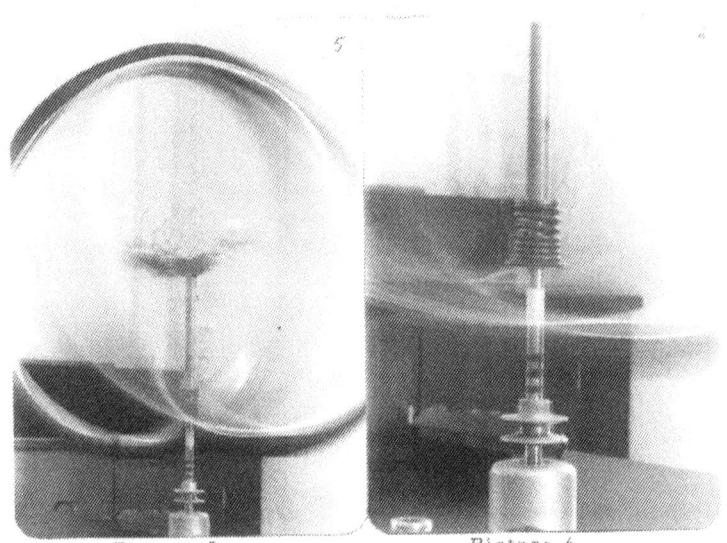

Picture 5

Picture 6

Picture 7

AMERICAN
JOURNAL
of PHYSICS

A Journal of the American Association of Physics Teachers

John S. Rigden, Editor
Philip B. James, Assistant Editor

Room 240 Benton Hall
University of Missouri - St. Louis
St. Louis, Missouri 63121 U.S.A.

April 22, 1981

Professor J. Parrilla-Calderon
Department of Math - Physics
Cayey University College
Cayey, PUERTO RICO 00633

Dear Professor Parrilla-Calderon:

Upon receipt of your revised manuscript "Force Generated
By Two Perpendicular Rotations II" we submitted your
revision to the original two referees and, in addition, to
a third neutral referee. Their reports are enclosed. On
the basis of these reviews, which are more or less self
explanatory, the American Journal of Physics will be unable
to accept your contribution for publication.

Sincerely,

Philip B. James
Assistant Editor

CC: Referees

106

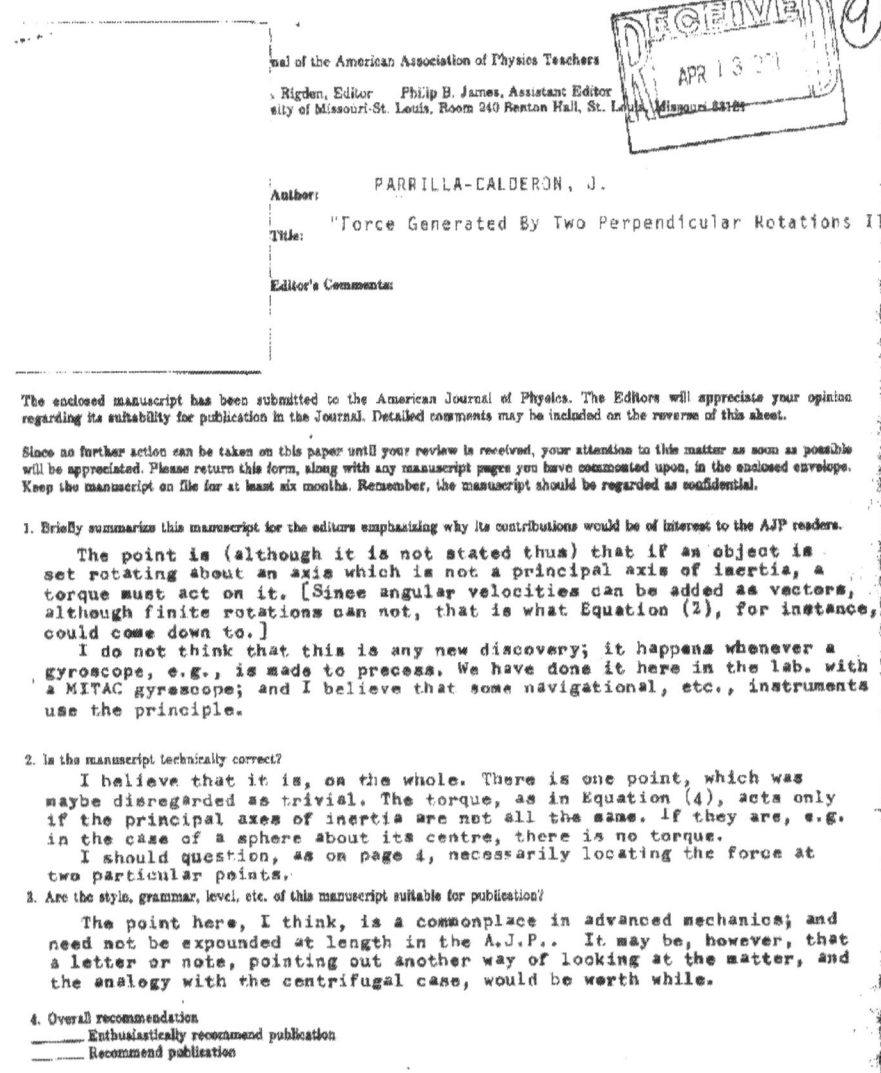

...nal of the American Association of Physics Teachers

. Rigden, Editor   Philip B. James, Assistant Editor
sity of Missouri-St. Louis, Room 240 Benton Hall, St. Louis, Missouri 83124

RECEIVED APR 1 3 (9)

Author: PARRILLA-CALDERON, J.

Title: "Force Generated By Two Perpendicular Rotations I..."

Editor's Comments:

The enclosed manuscript has been submitted to the American Journal of Physics. The Editors will appreciate your opinion regarding its suitability for publication in the Journal. Detailed comments may be included on the reverse of this sheet.

Since no further action can be taken on this paper until your review is received, your attention to this matter as soon as possible will be appreciated. Please return this form, along with any manuscript pages you have commented upon, in the enclosed envelope. Keep the manuscript on file for at least six months. Remember, the manuscript should be regarded as confidential.

1. Briefly summarize this manuscript for the editors emphasizing why its contributions would be of interest to the AJP readers.

The point is (although it is not stated thus) that if an object is set rotating about an axis which is not a principal axis of inertia, a torque must act on it. [Since angular velocities can be added as vectors, although finite rotations can not, that is what Equation (2), for instance, could come down to.]

I do not think that this is any new discovery; it happens whenever a gyroscope, e.g., is made to precess. We have done it here in the lab. with a MITAC gyroscope; and I believe that some navigational, etc., instruments use the principle.

2. Is the manuscript technically correct?

I believe that it is, on the whole. There is one point, which was maybe disregarded as trivial. The torque, as in Equation (4), acts only if the principal axes of inertia are not all the same. If they are, e.g. in the case of a sphere about its centre, there is no torque.

I should question, as on page 4, necessarily locating the force at two particular points.

3. Are the style, grammar, level, etc. of this manuscript suitable for publication?

The point here, I think, is a commonplace in advanced mechanics; and need not be expounded at length in the A.J.P.. It may be, however, that a letter or note, pointing out another way of looking at the matter, and the analogy with the centrifugal case, would be worth while.

4. Overall recommendation
_____ Enthusiastically recommend publication
_____ Recommend publication

_____ Can be published with revision
         _____ I should see the revision
         _____ I don't need to see the revision
__X__ Reject the manuscript (Reject as an article; a note, as suggested, might be
                                            worth while.)

ABOUT THE AUTHOR

The author is a retired Professor of Physics and researcher who also served as Chairman of the Department of Physics and Dean of Science at the Catholic University of Puerto Rico; Chairman of the Department of Mathematics and Physics at the University of Puerto Rico, Cayey Campus. Has done theoretical and experimental research on relativity theory, space-time physics, gyroscopic motion and nuclear radiation hazards. He also served one year as part-time Physics Lecturer at the City College of New York.

www.ingramcontent.com/pod-product-compliance
Lightning Source LLC
Chambersburg PA
CBHW022015170526
45157CB00003B/1251